THE
WORLD
IN
PERSPECTIVE

THE

WORLD

IN

PERSPECTIVE

A DIRECTORY OF WORLD MAP PROJECTIONS

Frank Canters and Hugo Decleir

JOHN WILEY & SONS

Chichester · New York · Brisbane · Toronto · Singapore

Copyright © 1989 by John Wiley & Sons Ltd
          Baffins Lane, Chichester
          West Sussex PO19 1UD, England

Other Wiley Editorial Offices

John Wiley & Sons, Inc., 605 Third Avenue,
New York, NY 10158-0012, USA

Jacaranda Wiley Ltd, G.P.O. Box 859, Brisbane,
Queensland 4001, Australia

John Wiley & Sons (Canada) Ltd, 22 Worcester Road,
Rexdale, Ontario M9W 1L1, Canada

John Wiley & Sons (SEA) Pte Ltd, 37 Jalan Pemimpin #05-04,
Block B, Union Industrial Building, Singapore 2057

Library of Congress Cataloging-in-Publication Data:

Canters, Frank.
    The world in perspective : a directory of world map projections /
  Frank Canters and Hugo Decleir.
        p.   cm.
    Bibliography: p.
    Includes index.
    ISBN 0 471 92147 5
    1. Map-projection.   I. Decleir, Hugo.   II. Title.
GA110.C34   1989
526'.8—dc20   89-14811
                                                    CIP

British Library Cataloguing in Publication Data:

Canters, Frank
    The world in perspective : a directory of
    world map projections.
    1. Map projections
    I. Title
    526.8

ISBN 0 471 92147 5

Printed in Great Britain by the Bath Press, Avon.

CONTENTS

PREFACE

'Maps organize space mathematically' Barry Lopez, *Arctic Dreams* (1987)

Maps, representing the world, were historically included in now famous map collections, printed atlases and geography books. They portrayed the configuration of continents and oceans and indicated the latest discoveries. As such, they fulfilled the basic curiosity of the philosopher, the statesman and the well-to-do bourgeois in the rapidly expanding universe of the sixteenth and seventeenth centuries. Looking at these maps must have made these people feel proud for the intrepidity of the discoverers and conquerors and given them a sense of security due to the richness and the unlimited space of the world. The making of these maps required both an artistic and technical skill and the possession of such artefacts could be considered as a status symbol for the privileged owner. The early printed maps are marvellously collected in a recent book by Rodney Shirley (1984).

Today the world map has ceased to arouse our own basic curiosity how the world looks like. This basic knowledge has been conveyed and imprinted by generations of school atlases, wall maps and encyclopedia. Images from space have replaced the world map in providing us with an authentic thrill and sense of discovery aroused by armchair exploration. However, the limits of space, the enclosure of continental and state boundaries of the present wildly overpopulated world, urges man to think about the effective organization of space on the surface of the globe. Representing the distribution of economic commodities, demographic facts, natural resources, strategic deployment, etc., are becoming daily routine in drawing up the 'state of the world'.

All too often the student of one or other thematic subject grasps the first map available to represent his data or to gather his information. But even when given it a more thorough thought it should be realized that the ideal projection does not exist and that each representation has its own drawbacks and distortions. Whereas the sphere occupies a well-defined space, the map projection organizes space and the well concepted map has to use effectively this organized information in the communication process between mapmaker and mapuser.

The construction of a map has been made simple by the use of computers. Given the basic formulas any one can—in principle—compute and plot a high-precision map of the world. The 'Directory' (Part II) of this book provides such formulas for 68 different views of the world. The description of each projection system is accompanied by one or two maps illustrating the deformation characteristics, allowing a quick evaluation of its merits and the effective use of the map for a particular purpose. All the maps shown have been produced and computed with the formulas given in the book and consequently tested for their accuracy and correctness.

The 'Directory' is preceded by a theoretical introduction about map projections. This provides in the necessary background information for the reader who is interested in making his own modification and/or change of aspect—in other words to allow for the construction of a personal projection system optimally adapted to a specific application.

Due to the systematic treatment of the map projections

ix

and the comprehensive quantitative evaluation and
mapping of the deformation characteristics the book
might also be used in a basic course on cartography.
Parts of the book were indeed originally concepted
for an undergraduate course in cartography offered by
the Department of Geography at the Free University
of Brussels (VUB). Other material has been collected
during a research project on automated cartography
which was partly sponsored by the Belgian National
Fund for Scientific Research through a research grant
to Frank Canters.

Brussels, Belgium   Frank Canters
August 1988   Hugo Decleir

Part I

PRINCIPLES
OF
MAP PROJECTIONS
AND THEIR
APPLICATION

THE
MATHEMATICAL THEORY
OF
MAP PROJECTIONS

1.1 The Nature of a Map Projection

In general mathematical terms a map projection can be defined as a one-to-one correspondence between points on a datum surface (the earth approximated by a sphere or ellipsoid) and points on a projection surface (a plane).

A system of parametric curves can be adopted on the datum surface as well as on the projection surface (Fig. 1.1). This allows us to refer to any point on these surfaces by means of coordinates. Denoting the coordinates on the datum surface as U, V and those on the projection surface as u, v the mathematical relationship between the coordinate systems can be expressed by

$$
\begin{aligned}
u &= f(U, V) \\
v &= g(U, V)
\end{aligned}
\tag{1}
$$

The functions f, g define a unique projection system and the nature of these functions determine the characteristics of the projection. A rational classification of projections must therefore be based on the nature of these functions. This subject will be dealt with later (Section 2.5).

For the moment it will suffice to point out that in the theory of map projection it is usually assumed that f and g are real, single valued, continuous and differentiable functions of u and v in a certain domain. Hence the Jacobian determinant $J(u, v)$ may not vanish:

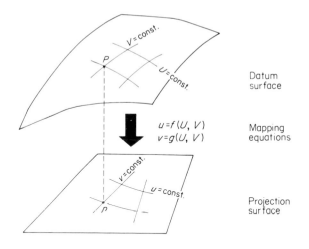

Fig. 1.1   Parametric curves on datum surface and projection surface.

$$
J(u, v) = \frac{\partial u}{\partial U} \frac{\partial v}{\partial V} - \frac{\partial v}{\partial U} \frac{\partial u}{\partial V} \neq 0
\tag{2}
$$

A global map is a two-dimensional representation of a three-dimensional solid body, the earth. The actual earth is flattened at the poles and is best approximated by an oblate ellipsoid. This is a figure generated by the rotation of an ellipse around its minor axis. However for the purpose of mapping the entire world onto a plane—which is the subject of this book—it will suffice to consider the earth as a perfect sphere with radius R_E = 6371 km.

A point on that sphere will be located by its spherical coordinates latitude ϕ and longitude λ. In the projection plane a point can be referenced by rectangular coordinates (x, y) or polar coordinates (r, θ). With rectangular coordinates the basic equations (1) and (2) become

$$x = f(\phi, \lambda)$$
$$y = g(\phi, \lambda) \qquad (3)$$

$$J(x, y) = \frac{\partial x}{\partial \phi}\frac{\partial y}{\partial \lambda} - \frac{\partial y}{\partial \phi}\frac{\partial x}{\partial \lambda} \neq 0 \qquad (4)$$

It should be remarked that the one-to-one correspondence between the earth and the map is never satisfied over the entire domain. Since the earth has a continuous surface and the map has a boundary some irregularities may occur at the edge of the map, e.g. the poles may be represented by lines. On some projections certain parts of the earth cannot be shown at all. Those points to which the one-to-one correspondence does not apply are called singular points.

The whole process of representing the earth on a map can now be schematized in three stages (Fig. 1.2):

1. First the size and shape of the earth is approximated by a mathematical figure, a datum surface, for the purpose of world maps a sphere with radius R_E = 6371 km.

2. Since a map is a small scale representation of the earth a scale reduction must take place that transforms the above mentioned spherical model into a smaller sphere which is called the 'generating globe'. The principal scale (also called nominal scale) of the map is then defined as the ratio of the radius of the generating globe R to the radius of the spherical model R_E,

$$S_N = \frac{R}{R_E} \qquad (5)$$

3. Finally, using relations (3), the map projection converts the generating globe into a map. The number of ways of accomplishing this step is infinite but, whatever the nature of the transformation may be, it always introduces some deformation. Nevertheless,

certain points or lines may be represented without any deformation. They will be called 'points or lines of no distortion'. It is evident that when trying to represent the earth on a plane the primary concern is to choose a map projection on which distortion is as small as possible. Accomplishing this necessitates an understanding of how the deformation takes place and how it is distributed over the entire map area. The mathematical theory of map projection provides us the necessary tools to study the deformation characteristics of an arbitrary projection. The fundamentals of this theory were laid down in the nineteenth century, first by Gauss and later by the French mathematician, Tissot.

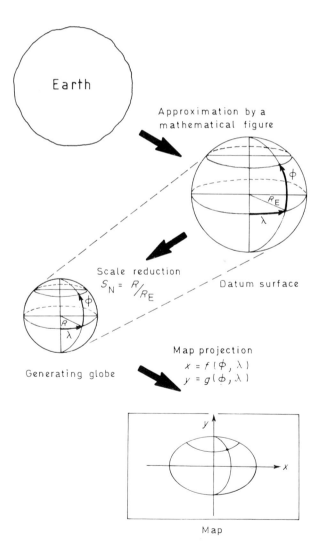

Fig. 1.2   Mapping process.

1.2 Some Elements of Gauss's Theory

Gauss's theory applies to the projection of any curved surface on another curved surface. For the more general cartographic treatment the reader is referred to Richardus and Adler (1971). As mentioned above it will for the purpose of this book suffice to approximate the earth by a reduced spherical model with radius R, called the generating globe. Figure 1.3a shows a line element $dS = PQ$ on that globe. The intersection of the parallels ϕ and $\phi + d\phi$ and the meridians λ and $\lambda + d\lambda$ through P and Q form an elementary quadrilateral. The diagonal line element dS makes an angle θ_m with the meridian and an angle θ_p with the parallel circle. Parallels and meridians form an orthogonal grid, hence $\theta = \theta_m + \theta_p = 90°$.

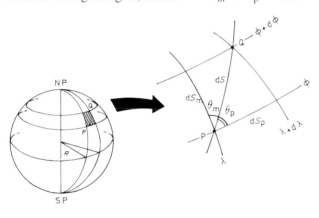

(a)

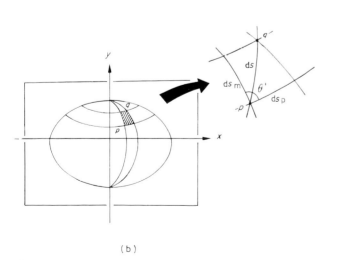

(b)

Fig. 1.3 Elementary quadrilateral on the generating globe (a), and on the projection surface (b).

If the quadrilateral is infinitely small Pythagoras' theorem applies:

$$dS^2 = dS^2_m + dS^2_p \qquad (6)$$

where $dS_m = R\,d\phi$ denotes an elementary arc along the meridian and $dS_p = R\cos\phi\,d\lambda$ an elementary arc along the parallel through P. The expression for the elementary length then becomes

$$dS = \sqrt{(R\,d\phi)^2 + (R\cos\phi\,d\lambda)^2} \qquad (7)$$

The azimuth A of this line element, equal to θ_m, is the clockwise direction from the north and given by

$$\tan A = \frac{R\cos\phi\,d\lambda}{R\,d\phi} = \cos\phi\frac{d\lambda}{d\phi} \qquad (8)$$

If this quadrilateral is mapped on a plane it will be distorted. Generally the direction and the length of the lines, as well as the angle between them, will be altered.

The projection of the points P and Q are p and q respectively, (Fig. 1.3b). Since the projected quadrilateral is also infinitely small the sides and diagonals can be considered as straight lines and Pythagoras' theorem applies again:

$$ds^2 = dx^2 + dy^2 \qquad (9)$$

By differentiating equation (3)

$$dx = \frac{\partial x}{\partial \phi}\,d\phi + \frac{\partial x}{\partial \lambda}\,d\lambda$$
$$dy = \frac{\partial y}{\partial \phi}\,d\phi + \frac{\partial y}{\partial \lambda}\,d\lambda \qquad (10)$$

Substituting in equation (9)

$$ds^2 = \left(\frac{\partial x}{\partial \phi}\,d\phi + \frac{\partial x}{\partial \lambda}\,d\lambda\right)^2 + \left(\frac{\partial y}{\partial \phi}\,d\phi + \frac{\partial y}{\partial \lambda}\,d\lambda\right)^2 \qquad (11)$$

and after rearranging terms one gets

$$ds^2 = E\,d\phi^2 + 2F\,d\phi\,d\lambda + G\,d\lambda^2 \qquad (12)$$

where E, F and G are the Gaussian fundamental quantities

$$E = \left(\frac{\partial x}{\partial \phi}\right)^2 + \left(\frac{\partial y}{\partial \phi}\right)^2$$

$$F = \frac{\partial x}{\partial \phi}\frac{\partial x}{\partial \lambda} + \frac{\partial y}{\partial \phi}\frac{\partial y}{\partial \lambda} \tag{13}$$

$$G = \left(\frac{\partial x}{\partial \lambda}\right)^2 + \left(\frac{\partial y}{\partial \lambda}\right)^2$$

Note that

$$\sqrt{EG - F^2} = \frac{\partial x}{\partial \lambda}\frac{\partial y}{\partial \phi} - \frac{\partial x}{\partial \phi}\frac{\partial y}{\partial \lambda} \tag{14}$$

is the Jacobian J as defined by equation (4).

1.2.1 Scale Distortion

The length or scale distortion m is defined by the ratio of the projected length ds over the original length dS on the generating globe:

$$m = \frac{ds}{dS}$$

$$= \sqrt{\frac{E\,d\phi^2 + 2F\,d\phi\,d\lambda + G\,d\lambda^2}{(R\,d\phi)^2 + (R\,\cos\phi\,d\lambda)^2}} \tag{15}$$

If at a certain point of the map $m = 1$ then the true scale S in that point equals the nominal scale. If m differs from 1, the projection has distorted the line element and the true scale is given by

$$S = S_{\mathrm{N}}m \tag{16}$$

The meaning of this cannot be overstressed. The reader should realize that the scale as indicated on a map is the nominal scale. This scale is applicable only on those points or along those lines where $m = 1$ (points or lines of no distortion).

Generally, however, the scale distortion is not one but varies from point to point and is different in every direction. This dependence on direction can be explicitly shown by substituting the expression for the azimuth equation (8) in equation (15):

$$m^2 = \frac{E}{R^2}\cos^2 A + \frac{G}{R^2\cos^2\phi}\sin^2 A$$

$$+ \frac{2F}{R^2\cos\phi}\sin A\,\cos A \tag{17}$$

1. For $A = 0$, $m = h$ represents the scale distortion along the meridian

$$h = \frac{ds_{\mathrm{m}}}{dS_{\mathrm{m}}} \qquad h = \frac{\sqrt{E}}{R}$$

$$= \frac{\sqrt{(\partial x/\partial \phi)^2 + (\partial y/\partial \phi)^2}}{R} \tag{18}$$

2. For $A = 90°$, $m = k$ represents the scale distortion along the parallel

$$k = \frac{ds_{\mathrm{p}}}{dS_{\mathrm{p}}} \qquad k = \frac{\sqrt{G}}{R\cos\phi}$$

$$= \frac{\sqrt{(\partial x/\partial \lambda)^2 + (\partial y/\partial \lambda)^2}}{R\cos\phi} \tag{19}$$

3. Since m varies with direction it is interesting to investigate in which directions the scale distortion attains its extrema. Using equations (18) and (19) and putting

$$p = \frac{2F}{R^2\cos\phi} \tag{20}$$

Equation (17) can be simplified to

$$m^2 = h^2\cos^2 A + k^2\sin^2 A + p\sin A\,\cos A \tag{21}$$

With the condition for an extremum

$$\frac{d}{dA}(m^2) = 0$$

it is found that

$$\tan 2A_{\mathrm{M}} = \frac{p}{h^2 - k^2} \tag{22}$$

indicating two orthogonal directions A_{M} and $A_{\mathrm{M}} + 90°$ where maximum and minimum scale distortion occur. These directions are called the *principal directions*.

1.2.2 Directions on the Map

The direction of a curve on the sphere is given by its azimuth A or the angles θ_{m} and θ_{p}. The direction of any curve in a point p on the map is given by the direction coefficient of the tangent in that point:

$$\tan \alpha = \frac{dy}{dx} = \frac{(\partial y/\partial \phi)d\phi + (\partial y/\partial \lambda)d\lambda}{(\partial x/\partial \phi)d\phi + (\partial x/\partial \lambda)d\lambda} \qquad (23)$$

The relation between the direction on the map and the direction on the sphere is obtained by substituting equation (8) in equation (23)

$$\tan \alpha = \frac{(\partial y/\partial \phi) \cos \phi \cos A + (\partial y/\partial \lambda) \sin A}{(\partial x/\partial \phi) \cos \phi \cos A + (\partial x/\partial \lambda) \sin A} \qquad (24)$$

1. Putting $A = 0$ gives the north direction or the direction α_m of the meridian

$$\tan \alpha_m = \frac{\partial y/\partial \phi}{\partial x/\partial \phi} \qquad (25)$$

2. Putting $A = 90°$ gives the east direction or the direction α_p of the parallel

$$\tan \alpha_p = \frac{\partial y/\partial \lambda}{\partial x/\partial \lambda} \qquad (26)$$

The angle of intersection θ' between parallels and meridians (Fig. 1.4) can now be calculated

$$\tan \theta' = \tan (\alpha_m - \alpha_p)$$
$$= \frac{\tan \alpha_m - \tan \alpha_p}{1 + \tan \alpha_m \tan \alpha_p} \qquad (27)$$

With the use of equations (25) and (26) and after substituting equations (13) and (14)

$$\tan \theta' = \frac{(\partial x/\partial \lambda)(\partial y/\partial \phi) - (\partial x/\partial \phi)(\partial y/\partial \lambda)}{(\partial x/\partial \phi)(\partial x/\partial \lambda) + (\partial y/\partial \phi)(\partial y/\partial \lambda)}$$
$$= \frac{\sqrt{EG - F^2}}{F} \qquad (28)$$

or alternatively

$$\sin \theta' = \sqrt{\frac{EG - F^2}{EG}} \qquad (29)$$

An interesting mapping technique becomes evident from equation (29). If one makes $F = 0$ then $\theta' = 90°$ which means that the orthogonality of parallels and meridians is preserved during mapping. Equation (22) indicates that in that case ($F = 0$) the principal directions are situated along parallels and meridians. This will become more evident when studying the indicatrix of Tissot.

1.2.3 Conformal Projections

Conformal projections are obtained when the scale distortion is independent from azimuth or is the same in every direction. Equation (17) shows that in addition to $F = 0$ one has to make

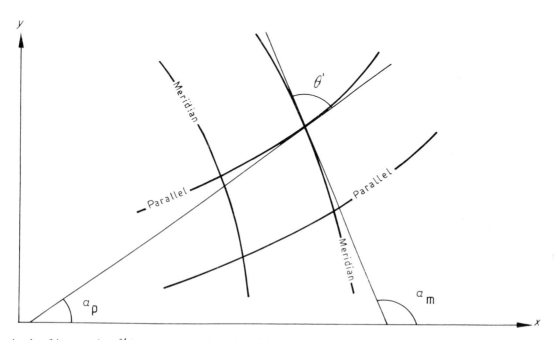

Fig. 1.4 Angle of intersection θ' between parallels and meridians in the mapping plane.

$$\frac{E}{R^2} = \frac{G}{R^2 \cos^2 \phi} \qquad (30)$$

In that case $m = h = k$. Due to the fact that the length distortion is the same in every direction there is no angular distortion hence the name 'conformal projection'.

1.2.4   Equal-area Projections

If during the mapping process elementary areas are preserved a projection is said to be equal-area or equivalent.

An elementary surface on the sphere is given by $dS_m \, dS_p = R^2 \cos \phi \, d\phi \, d\lambda$, on the image plane by $ds_m \, ds_p \sin \theta'$.

Hence the surface distortion is defined as

$$\sigma = \frac{ds_m \, ds_p \, \sin \theta'}{dS_m \, dS_p} = hk \, \sin \theta' \qquad (31)$$

and with equations (18), (19) and (29)

$$\sigma = \frac{\sqrt{EG - F^2}}{R^2 \cos \phi} \qquad (32)$$

The condition for an equal-area projection ($\sigma = 1$) is therefore given by

$$\sqrt{EG - F^2} = R^2 \cos \phi \qquad (33)$$

1.3   Tissot's Indicatrix

1.3.1   The Concept

If one calculates, in the point P on the sphere, the scale distortion m for each direction and then plots the computed values in the corresponding directions from the point p on the map, the obtained locus of points is called the indicatrix of Tissot. By choosing $dS = 1$ this indicatrix is also the representation in the plane of an elementary circle with radius-1 on the sphere. It will now be shown that this indicatrix is an ellipse.

The system of linear equations (10) can be solved for $d\phi$ and $d\lambda$:

$$d\phi = \frac{(\partial x / \partial \lambda) \, dy - (\partial y / \partial \lambda) \, dx}{\sqrt{EG - F^2}}$$

$$d\lambda = \frac{(\partial y / \partial \phi) \, dx - (\partial x / \partial \phi) \, dy}{\sqrt{EG - F^2}} \qquad (34)$$

With the introduction of equation (34) in equation (7) the expression for the radius of the elementary circle becomes

$$dS^2 = \frac{R^2 [(\partial x / \partial \lambda) \, dy - (\partial y / \partial \lambda) \, dx]^2}{EG - F^2}$$
$$+ \frac{R^2 \cos^2 \phi [(\partial y / \partial \phi) \, dx - (\partial x / \partial \phi) \, dy]^2}{EG - F^2} \qquad (35)$$

This expression can be further developed with the use of the following abbreviations

$$A = R^2 \left(\frac{\partial y}{\partial \lambda}\right)^2 + R^2 \cos^2 \phi \left(\frac{\partial y}{\partial \phi}\right)^2$$

$$B = -2 \left(R^2 \frac{\partial x}{\partial \lambda} \frac{\partial y}{\partial \lambda} + R^2 \cos^2 \phi \frac{\partial x}{\partial \phi} \frac{\partial y}{\partial \phi}\right) \qquad (36)$$

$$C = R^2 \left(\frac{\partial x}{\partial \lambda}\right)^2 + R^2 \cos^2 \phi \left(\frac{\partial x}{\partial \phi}\right)^2$$

$$D = -(EG - F^2)$$

to

$$A \left(\frac{dx}{dS}\right)^2 + B \left(\frac{dx}{dS} \frac{dy}{dS}\right) + C \left(\frac{dy}{dS}\right)^2 + D = 0 \qquad (37)$$

Putting $dS = 1$, the second degree equation (37) shows that the elementary circle is transformed on the map into an ellipse. The classical expression for the ellipse is obtained by rotating the coordinate systems so that the x, y term in equation (37) disappears:

$$\frac{d\bar{x}^2}{[(EG - F^2)/A]} + \frac{d\bar{y}^2}{[(EG - F^2)/C]} = 1 \qquad (38)$$

The major and minor diameters of the ellipse are now lying along the $\bar{x}, \bar{y}$ axis of the rotated coordinate system (Fig. 1.5). The semidiameters a and b of this ellipse correspond with maximum and minimum scale distortion respectively and hence with the principal directions.

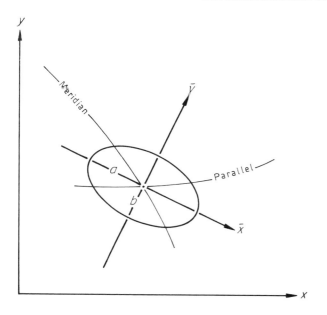

Fig. 1.5   Indicatrix of Tissot.

1.3.2   Analysis of the Deformational Characteristics by means of the Indicatrix of Tissot

From the foregoing it is clear that the principal directions remain orthogonal after projection. The analysis of the infinitesimal deformation characteristics is greatly simplified by choosing both in the tangent plane

in the point P of the sphere and in the mapping plane a Cartesian coordinate system oriented along those principal axes of deformation.

Figure 1.6a shows in the tangent plane on the sphere such a X-, Y-coordinate system and a point Q on the elementary circle with radius 1 and centre P. The elementary line element $PQ = 1$ makes an angle α with the X-axis.

Figure 1.6b shows the projection of the elementary circle in the mapping plane. The point Q is mapped as q and is lying now on an ellipse with semidiameters a and b directed along the x, y coordinate axes. The angle α of the original direction is now transformed to α'. By convention the x-axis is chosen along the principal direction with maximum scale distortion while the y-axis corresponds with minimum distortion.

The equation of the indicatrix is

$$\frac{x^2}{a^2} + \frac{y^2}{b^2} = 1 \quad \text{with } a \geq b \qquad (39)$$

where the semidiameters a and b correspond with the extremal scale distortions. They are also called the principal scale factors. It will now be shown how all the deformation characteristics can be related to these principal scale factors a and b. This means that visual inspection of the indicatrix will give a quick and complete information regarding the deformational characteristics in a certain point of the map.

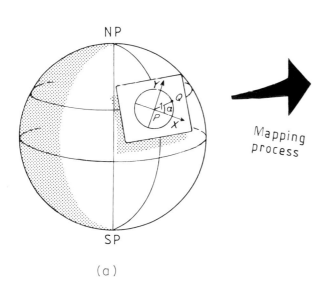

(a)

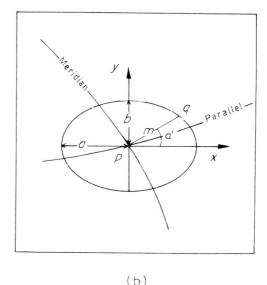

(b)

Fig. 1.6   Elementary circle and principal axes in the tangent plane on the sphere (a), and corresponding ellipse in the mapping plane (b).

Length distortion

The length distortion along the principal axes has already
been defined as a and b. From the definition of the scale
distortion (Section 1.2.1.) and from Fig. 1.6 it is found
that

$$a = \left(\frac{ds}{dS}\right)_x = \frac{m \cos \alpha'}{\cos \alpha}$$

and   (40)

$$b = \left(\frac{ds}{dS}\right)_y = \frac{m \sin \alpha'}{\sin \alpha}$$

or

$$a \cos \alpha = m \cos \alpha'$$

and   (41)

$$b \sin \alpha = m \sin \alpha'$$

It follows that

$$m^2 = a^2 \cos^2 \alpha + b^2 \sin^2 \alpha \qquad (42)$$

which expresses the length distortion in any direction as
a function of the original direction α and the principal
scale factors a and b.

Parallels and meridians are orthogonal on the sphere.
If α is the direction of the parallel with respect to the
principal X-axis, then $\beta = \alpha + \pi/2$ is the direction of
the meridian (Fig. 1.7a). The scale distortions along
parallels and meridians, denoted k and h respectively,
are therefore given by

$$k^2 = a^2 \cos^2 \alpha + b^2 \sin^2 \alpha \qquad (43)$$

$$h^2 = a^2 \cos^2 \beta + b^2 \sin^2 \beta \qquad (44)$$

Due to the orthogonality the latter equation can also be
written as

$$h^2 = a^2 \sin^2 \alpha + b^2 \cos^2 \alpha \qquad (45)$$

Equations (43) and (45) combine to

$$h^2 + k^2 = a^2 + b^2 \qquad (46)$$

an expression which will be used later (Section 1.3.3).

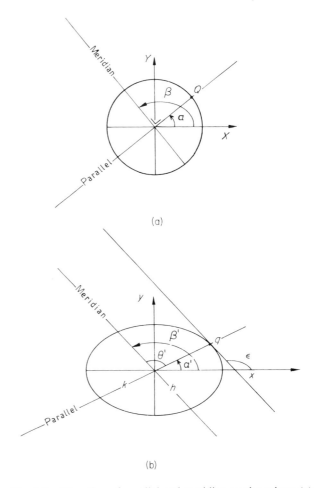

Fig. 1.7   Direction of parallel and meridian on the sphere (a),
and in the mapping plane (b).

Angular distortion

The direction of a line element with respect to the
principal direction is given by $\tan \alpha = Y/X$ (Fig. 1.6a).
After projection the direction of the line element
becomes (Fig. 1.6b)

$$\tan \alpha' = \frac{y}{x} = \frac{m \sin \alpha'}{m \cos \alpha'}$$

$$= \frac{b \sin \alpha}{a \cos \alpha} = \frac{b}{a} \tan \alpha \qquad (47)$$

The change in direction can be expressed by means of

$$\tan (\alpha - \alpha') = \frac{\tan \alpha - \tan \alpha'}{1 + \tan \alpha \tan \alpha'} \qquad (48)$$

and with the help of equation (47)

$$\tan(\alpha - \alpha') = \frac{a - b}{(a/\tan\alpha) + b\tan\alpha} \qquad (49)$$

it shows that both for $\alpha = 0$ and $\alpha = 90°$, $\alpha - \alpha' = 0$ which confirms the fact that the principal axes remain orthogonal during the mapping process. The maximum change in direction occurs when the denominator in equation (49) reaches a minimum. This occurs for a value α_{max} given by

$$\tan\alpha_{max} = \pm\sqrt{\frac{a}{b}} \qquad (50)$$

The corresponding value α'_{max} is found by introducing equation (50) in equation (47)

$$\tan\alpha'_{max} = \pm\sqrt{\frac{b}{a}} \qquad (51)$$

Using equations (50) and (51) the maximum change in direction is written as

$$\tan(\alpha - \alpha')_{max} = \tan\Omega = \frac{a - b}{2\sqrt{ab}} \qquad (52)$$

or alternatively

$$\sin\Omega = \frac{a - b}{a + b} \qquad (53)$$

From equations (50) and (51) it can easily be verified that $(\alpha_{max} + \alpha'_{max})/2 = (2i - 1)\pi/4$ where i denotes the quadrant number. This means that the directions α_{max} and α'_{max} mirror about the bisectrix (Fig. 1.8). Expression (49) indicates then that the change in direction $\alpha - \alpha'$ has an opposite sign in each two adjacent quadrants. As shown this leads to a maximum angular distortion 2Ω which occurs for the angle subtended by the directions of maximum distortion in two adjacent quadrants and given by

$$2\Omega = 2\arcsin\frac{a - b}{a + b} \qquad (54)$$

If $2\Omega = 0$ for every point on the map no angular deformation will occur and the projection is called conformal. The property of conformality is now given by $a = b$ which implies that Tissot's indicatrix is a circle and the scale distortion is equal in all directions. The condition $a = b$ is identical with the previously given

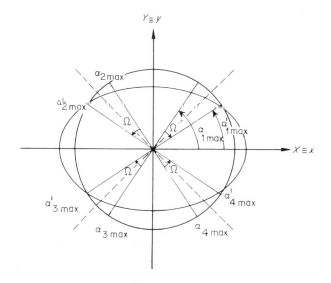

Fig. 1.8 Superposition of elementary circle (generating globe) on the indicatrix (mapping plane) showing maximum change in direction Ω in the four quadrants, numbered 1–4.

condition for conformality $h = k$ and $\theta' = \pi/2$ (Section 1.2.3.). Figure 1.9 shows the grid of the cylindrical conformal projection (Mercator). In every grid point the indicatrix is represented as a circle. Note, however, that the area of the circles increases with increasing latitude. On the equator, which is a line of no distortion, the indicatrices are represented as elementary circles with unit radius.

Conformal projections are used for topographic, military and navigational purposes.

Areal distortion

The areal distortion is easily found by dividing the surface of the indicatrix by the corresponding surface of the circle with radius 1.

$$\sigma = \frac{\pi a b}{\pi} = ab \qquad (55)$$

Equations (31) and (55) combine to

$$ab = hk\sin\theta' \qquad (56)$$

In the special case where $\sigma = ab = 1$ throughout the entire area of the map the projection is called an equal-area or equivalent projection. It is clear that conformality ($a = b$) and equivalence ($ab = 1$) are two mutually exclusive properties. Otherwise no deformation at all

MERCATOR' S PROJECTION

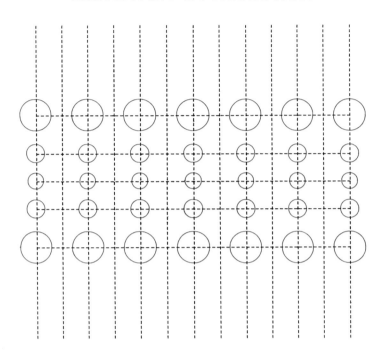

0   5000 KM

Fig. 1.9   Indicatrices on Mercator's conformal grid.

would occur, which is mathematically impossible since a sphere cannot be developed in a plane.

Figure 1.10 shows the indicatrices for the cylindrical equal-area projection with two standard parallels (Behrmann). They are represented as ellipses with an area equal to the area of the elementary circle with unit radius. Along the two standard parallels, which are lines of no distortion, the indicatrices become elementary circles.

The special property of equivalence is often required for small scale maps that show statistical distributions.

The transformed angle θ'

For $\theta' = \pi/2$ parallels and meridians remain orthogonal during the mapping process. In that case equation (56) shows that

$$h = a \text{ or } b \quad \text{and} \quad k = b \text{ or } a$$

In other words the orthogonality is obtained at the expense of a maximum difference in scale distortion.

The projections of Figs. 1.9 and 1.10 are examples of this mapping technique. However, generally $\theta' \neq \pi/2$ and parallels and meridians are not situated along the principal axes (Fig. 1.7b). The direction of the meridian on the map is then given by (cf. equation 47)

$$\tan \beta' = \frac{b}{a} \tan \beta = -\frac{b}{a} \cotan \alpha \qquad (57)$$

The direction coefficient of the tangent line in the point q (Fig. 1.7b)—intersection of the parallel with the indicatrix—is found by differentiation of the equation of the ellipse

$$\tan \varepsilon = \frac{dy}{dx} = -\frac{x}{y} \frac{b^2}{a^2} \qquad (58)$$

Remembering that $x = m \cos \alpha' = a \cos \alpha$ and $y = m \sin \alpha' = b \sin \alpha$ one finds

$$\tan \varepsilon = -\frac{b}{a} \cotan \alpha = \tan \beta' \qquad (59)$$

Since both the meridian and the tangent line in the point q of the indicatrix have the same direction, it follows

CYLINDRICAL EQUAL—AREA PROJECTION
WITH TWO STANDARD PARALLELS
(BEHRMANN)

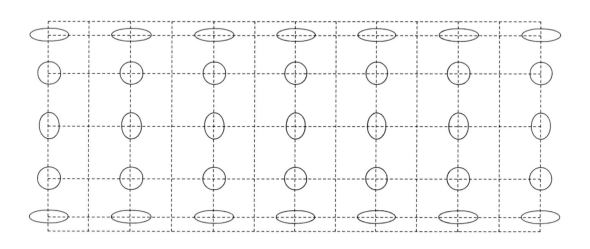

0   5000 KM

Fig. 1.10   Indicatrices on Behrmann's equal-area grid.

that parallels and meridians, subtending an angle θ', are represented as conjugate diameters on the indicatrix.

With equations (57) and (47) the transformed angle $\theta' = \beta' - \alpha'$ can be written as

$$\tan \theta' = \frac{2ab}{\sin 2\alpha(b^2 - a^2)} \tag{60}$$

With increasing α the transformed angle deviates more and more from $90°$ while the difference between h and k diminishes. For $\alpha = 45° \theta'$ attains its greatest value

$$\tan \theta' = \frac{2ab}{b^2 - a^2} \tag{61}$$

For that value of θ' it can easily be shown that

$$h = k = \sqrt{\frac{a^2 + b^2}{2}} \tag{62}$$

Figure 1.11 shows the indicatrices for the Winkel–Tripel projection where generally $\theta' \neq \pi/2$. Note how the scale distortions h and k are now represented as conjugate semi-diameters of the ellipse.

1.3.3   Conclusion

The projection of a sphere on a plane always introduces deformation. As a result of this deformation the scale varies from point to point and is generally different in every direction. An infinitely small circle around a point on the globe will be mapped as an ellipse, the so-called indicatrix of Tissot. This indicatrix describes the deformation characteristics of the projection in this point. Scale distortion reaches a maximum and minimum in the direction of the main axes of the indicatrix. When these extreme values are known it is possible to calculate areal and angular distortion as well as all other information about distortion. The maximum and minimum scale distortion can be obtained by solving the following set of equations for a and b

$$a^2 + b^2 = h^2 + k^2 \tag{46}$$

$$ab = hk \sin \theta' \tag{56}$$

where h, k and θ' are first calculated from the partial derivations of x and y with respect to ϕ and λ, equations (18), (19) and (28).

WINKEL—TRIPEL PROJECTION

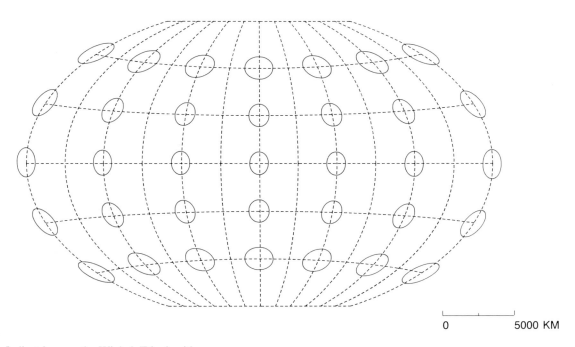

0   5000 KM

Fig. 1.11   Indicatrices on the Winkel–Tripel grid.

Maximum angular deformation and areal distortion are then easily obtained by

$$2\Omega \; = \; 2\arcsin\frac{a-b}{a+b} \qquad (54)$$

$$\sigma \; = \; ab \qquad (55)$$

The study of the deformation characteristics of a projection plays an important role in the process of selecting a suitable projection for a particular purpose. This subject will be treated later (Section 3.1). First an attempt will be made to classify projections in terms of their deformation patterns as these strongly influence the visual appearance of the map.

2

THE
CLASSIFICATION
OF
MAP PROJECTIONS

From the definition of a map projection it is evident that the number of possible representations of the earth on a plane is unlimited. For world maps alone hundreds of projections have been devised although many of them have little practical significance. Also many so called new projections are merely slight modifications of already existing ones while, due to the bad habit of designating a projection by the name of its author, the relation of the modification to the parent projection often becomes indistinct. Therefore, in order to gain more insight in the large variety of map projections and the relations that may exist among them, a rational classification seems necessary. Such a classification is, moreover, of considerable practical value once a suitable projection has to be selected for a particular purpose, especially if the criteria for the grouping coincide with those used in the selection process. These criteria involve both visual and deformational considerations. Once a classification has been set up it becomes possible to adapt the nomenclature of projections to it. This allows a projection to be described in terms of its visual appearance and of some other properties and will therefore be less confusing than naming the projection according to its 'supposed' author.

Before deepening the subject of classification and nomenclature the following concepts will be introduced: patterns of distortion, the aspect of a map projection, special properties and transformations of a map projection.

2.1 Patterns of Distortion

The patterns of distortion show the distribution of the various deformation characteristics involved in the mapping process. An obvious method for representing the patterns of distortion is the construction of isolines for certain deformation parameters. In practice this will generally be restricted to the maximum angular deformation 2Ω and the areal distortion σ (see Directory (part II) for examples).

A comparison of the distortion patterns for different projections and for different aspects of the same projection ('aspects' of a projection are explained in the next section) shows that a typical distortion pattern is related to the appearance of the graticule in its normal aspect. These patterns will therefore serve as a basis for classification.

It has already been said that for most projections no angular and no areal distortion takes place along certain lines or at certain points on the map. The traditional geometric classification of projections into an azimuthal, a cylindrical and a conical class can be brought back to the configuration and location of these lines or points of zero distortion.

1. *Azimuthal projections* have a central point of zero distortion. The distortion parameters increase radially from this central point leading to a distortion pattern of concentric circular isolines. The characteristic grid pattern of an azimuthal projection is likewise circular: in the normal aspect meridians are concurrent straight lines and parallels concentric circles with center at the pole.

2. *Cylindrical projections* have a single line of zero distortion, also called standard line. This line corresponds to a great circle on the globe and is represented on the map by a straight line. The distortion parameters increase in a direction perpendicular to and away from the standard line. The distortion pattern thus consists of rectilinear isolines parallel and symmetrical to the standard line. The characteristic grid of a cylindrical projection is rectangular. In the normal aspect parallels and meridians form an orthogonal rectilinear grid. The poles are then represented as straight lines equal in length to the equator which coincides with the line of zero distortion.

3. *Conical projections* have also a single line of zero distortion but corresponding now to a small circle on the globe. This small circle is represented on the map as a circular arc. The distortion parameters increase in a direction perpendicular to and away from the standard line so that the distortion isolines are also circular arcs concentric with the standard line. The characteristic grid of a conical projection is fan shaped. In the normal aspect meridians are concurrent straight lines, parallels are concentric circular arcs. The parallel which coincides with the line of zero distortion is called standard parallel. The center of the circular arcs representing the parallels is not necessarily the pole. In general the pole is also shown as a circular arc.

The names 'cylindrical' and 'conical' projections follow from the resemblance of their distortion patterns with that of a purely geometrical projection of the spherical earth on a tangent cylinder or cone respectively. The parallel of contact on the globe is then mapped without any deformation and corresponds with the line of zero distortion on the map (Fig. 2.1b,c). In a similar way 'azimuthal' projections have a distortion pattern resembling that of a perspective projection of the globe on a plane tangent to the earth. In this case the point of contact coincides with the point of no distortion on the map (Fig. 2.1a).

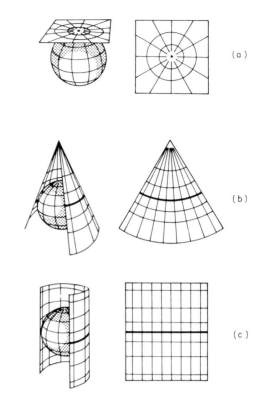

Fig. 2.1   Geometrical projection of the sphere on (a) a tangent cylinder, (b) a tangent cone and (c) a tangent plane.

It should be added that by pure geometrical construction it is also possible to project the sphere on a cylinder, cone or plane which intersects the globe. The distortion pattern will then show two lines of zero distortion (instead of one) for the cylindrical and conical projections and one standard circle (instead of a point) for the azimuthal projections.

Generalizing, the terms cylindrical, conical or azimuthal projection will also be used whenever the distortion patterns resemble those generated by perspective projection, even if they are generated by a purely analytical transformation.

2.2   The Aspect of a Map Projection

As discussed in the previous section, the distortion pattern of a projection shows how distortion varies from place to place. The least distortion occurs of course around the line(s) or point(s) of zero distortion. This means that by a careful repositioning of the distortion pattern of a projection with respect to the globe it is possible to place an arbitrary region of interest in the least distorted area of the map. Such a repositioning is denoted as a change of aspect and is accomplished by a transformation of the map graticule. It has no effect on the nature of the distortion pattern itself, only the appearance of graticule and therefore the continental configuration is altered.

In the normal aspect of a projection the geographical pole P can be considered as the mathematical pole of a reference great circle Q which is called the equator. Reference circle and pole determine the normal geographical latitude ϕ and longitude λ (Fig. 2.2). In general however a projection can be mathematically developed with respect to a reference system determined by a great circle Q', which is oblique to the equator. The associated mathematical pole T therefore does not coincide any longer with one of the geographical poles. In this new reference system a point is referred to by its transformed latitude ϕ' and longitude λ' (Fig. 2.2). The transformation of the normal geographical coordinates ϕ, λ of a point M to the new coordinates ϕ', λ' of the oblique system is accomplished by solving the spherical triangle PTM.

$$\sin \phi' = \sin \phi \sin \phi_T$$
$$+ \cos \phi \cos \phi_T \cos(\lambda - \lambda_T) \quad (63)$$

$$\sin \lambda' = \frac{\sin(\lambda - \lambda_T) \cos \phi}{\cos \phi'} \quad (64)$$

$$\cos \lambda' = \frac{\cos \phi \sin \phi_T \cos(\lambda - \lambda_T) - \sin \phi \cos \phi_T}{\cos \phi'} \quad (65)$$

The transformed latitude ϕ' is unambiguously determined by equation (63). For the transformed longitude λ' which has a range from $-180°$ to $180°$ quadrant adjustment is necessary. The calculation of $\sin \lambda'$ and $\cos \lambda'$, given by equations (64) and (65), makes identification of the quadrant possible.

The use of modern computer equipment makes the change in aspect relatively easy and allows for an accurate drawing of the graticule. As an example Fig. 2.3 shows the azimuthal equidistant projection for gradually decreasing latitude ϕ_T of the mathematical pole T. For $\phi_T = 90°$ the normal aspect of the projection is obtained. It is seen that the pole of the projection (which on an azimuthal projection is the point of zero distortion) coincides with the geographical pole. By altering the position of this projection pole T the graticule becomes more complex while the distortion pattern remains unchanged. The special case where the pole is situated on the equator ($\phi_T = 0°$) is referred

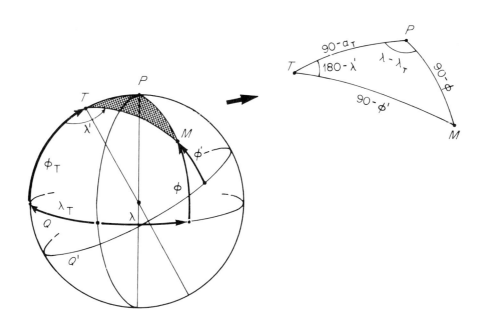

Fig. 2.2   Coordinate transformation for oblique aspect.

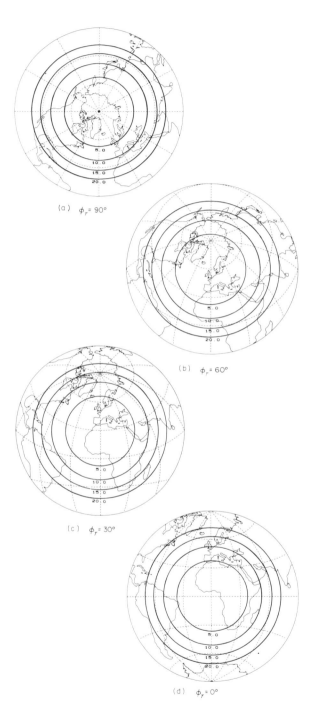

(a) $\phi_r = 90°$

(b) $\phi_r = 60°$

(c) $\phi_r = 30°$

(d) $\phi_r = 0°$

Fig. 2.3   Lines of equal angular distortion for different aspects of the azimuthal equidistant projection: (a) polar aspect, $\phi_T = 90°$; (b) oblique aspect, $\phi_T = 60°$; (c) oblique aspect, $\phi_T = 30°$; (d) transverse aspect, $\phi_T = 0°$.

to as the transverse aspect of a projection. The three aspects of a projection (normal, transverse and oblique) each correspond to a particular relationship between graticule and distortion pattern of a projection. This will be illustrated by studying these aspects for three classes of projections: the azimuthal, the cylindrical and the pseudocylindrical class. The azimuthal and cylindrical class have already been described in the previous section. The pseudocylindrical class groups the projections that represent in the normal aspect the parallels by straight parallel lines and the meridians by concurrent curves. This class of projections is extremely important for world maps (most of the projections dealt with in this book belong to this class). The distortion pattern of these projections has only two axes of symmetry. Therefore pseudocylindrical projections allow us to describe the different aspects of a projection in a more general context.

2.2.1   The Normal Aspect

In the normal aspect of a projection there is a direct relationship between the graticule and the distortion pattern. This means that the grid has the same axes of symmetry as the distortion pattern.

For both the azimuthal and the cylindrical projections the distortion patterns were described in the previous section. In the normal aspect the lines of equal distortion coincide with the parallels. For the azimuthal projection the point of zero distortion corresponds with one of the geographical poles. Therefore the normal aspect of an azimuthal projection is also called the polar aspect (Fig. 2.3a). For the cylindrical projection the line of zero distortion corresponds with the equator. Hence the normal aspect of this class of projections is known as the equatorial aspect (Fig. 2.4a).

The distortion pattern of a pseudocylindrical projection has only two axes of symmetry which correspond in the normal aspect with the equator and the central meridian. Numerous examples of normal aspect pseudocylindrical projections are shown in the Directory.

2.2.2   The Transverse Aspect

In the transverse aspect the distortion pattern is rotated by 90°. As can be seen from Fig. 2.3d the point of zero distortion of an azimuthal projection is now situated on the equator. Therefore the transverse aspect of this class of projections is often referred to as the equatorial aspect. The lines of equal distortion do not correspond

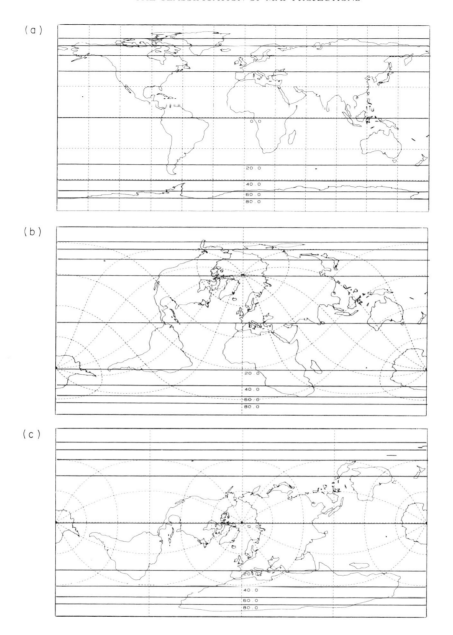

Fig. 2.4   Lines of equal angular distortion for different aspects of the cylindrical equidistant projection: (a) equatorial aspect, $\phi_T = 90°$; (b) oblique aspect, $\phi_T = 45°$; (c) transverse aspect, $\phi_T = 0°$.

any longer with the parallels of latitude. The level of symmetry has clearly decreased with respect to the normal aspect: the graticule is now symmetrical both about the central meridian and the equator. For the cylindrical projections the central meridian becomes the line of zero distortion in the transverse aspect. Figure 2.4c shows how the graticule now consists of curved meridians and parallels. Only the equator, the central meridian and the meridians that make an angle of 90°

with it are represented as straight lines which also serve as axes of symmetry.

The transverse aspect of a pseudocylindrical projection can be generated from the normal aspect in two ways that both lead to a different appearance of the graticule:

1. The distortion pattern is rotated about 90° in such a way that one of the geographical poles coincides

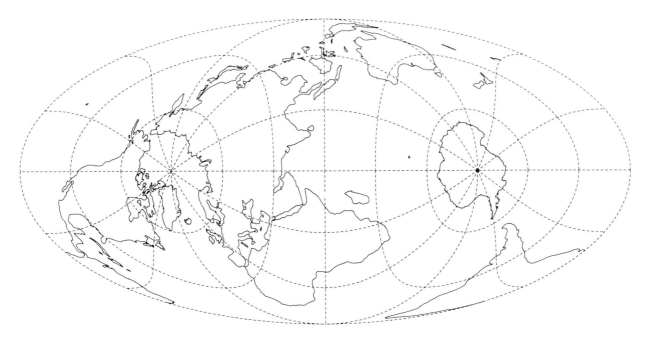

Fig. 2.5 Mapping process. Transverse aspect Mollweide projection with longer axis, 60°E, 120°W.

with the center of the projection. The two axes of symmetry of the distortion pattern are now lying along two perpendicular meridians. Thus, although the graticule has become much more complicated it has preserved its symmetry about two axes (e.g. transverse aspect Mollweide projection (see Directory, p.102)).

2. The distortion pattern is rotated about 90° without altering the position of the center of the projection. The longer axis has now become the shorter one and vice versa. In this case also meridians and parallels have become much more complicated, but nevertheless the two-sided symmetry has remained (Fig. 2.5).

2.2.3 The Oblique Aspect

In the oblique aspect of a projection the center of the distortion pattern can be located anywhere on the globe and the axes of symmetry can have every possible direction. The normal and transverse aspect can be considered as a limiting case of the oblique aspect.

The oblique aspect of the azimuthal projection is characterized by a point of zero distortion that is neither situated on one of the geographical poles nor on the equator. The direct relationship between distortion pattern and graticule has disappeared: all meridians and

parallels are curved, except for the central meridian which is also the single axis of symmetry (Fig. 2.3b,c).

In the oblique aspect of the cylindrical projection the line of zero distortion coincides with an arbitrary great circle on the globe. The graticule is symmetrical about one axis only and this corresponds with the meridional great circle on which the line of zero distortion reaches its vertex (Fig. 2.4b).

The oblique aspect of the pseudocylindrical projection can be generated in two ways:

1. One of the two axes of symmetry of the distortion pattern coincides with a central meridian which is the single axis of symmetry of the graticule (e.g. Atlantis projection [see Directory, p.104]).
2. None of the two axes of the distortion pattern coincides with a meridian. This special case of the oblique aspect is often called 'skew oblique' and the corresponding maps are referred to as 'skew oblique projections'. In general the graticule of those maps shows no symmetry (Fig. 2.6).

Since a change of aspect does not influence the distortion pattern of a projection, only the appearance of the graticule and of the continents will change. Different aspects of the same projection are therefore not to be considered as different projections. In the past many cartographers generated particular aspects of existing projections and gave distinct names to them.

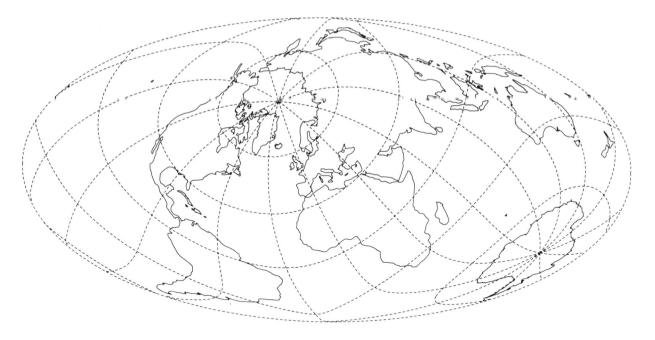

Fig. 2.6 Skew oblique Hammer–Aitoff projection.

This only leads to confusion. All aspects of a projection satisfy the same functional relationships so there is absolutely no need to give different names to each aspect of a projection.

2.3 Special Properties of a Map Projection

The distortion patterns allow us to subdivide all projections into a number of classes. Projections belonging to the same class may have certain properties that make them useful for a particular purpose. The best known special properties are equivalence and conformality: they arise from a special relationship between the principal scale factors which is maintained throughout the whole map area. Other special properties include equidistance, the property of representing all circles on the sphere as circles on the map, the property of representing the orthodrome or loxodrome as a straight line, etc.

A projection is said to be equivalent (equal-area) if the product of maximum and minimum scale distortion equals unity in every point on the map (except in the singular points). This means that no distortion of areas takes place. It should be realized that although this concept of equivalence was originally defined on an infinitely small area around a point, it may also be applied to areas of finite size. This implies for instance

that the different continents on an equal-area map are represented in correct proportion.

A projection is said to be conformal if the principal scale factors are identical (or the maximum angular distortion is zero) in every point on the map. The concept of conformality—which was also defined on an infinitely small scale—does, however, not guarantee a good representation of large shapes. Due to the great variability in scale from point to point significant deformation does occur when considering large areas. Conformal projections are therefore less suited for small-scale cartography.

The property of equidistance implies that the scale distortion equals unity. Since this condition can only be satisfied in a particular direction, e.g. along meridians ($h = 1$) or along parallels ($k = 1$), equidistant projections in the restricted sense do not exist. Generally however projections which maintain the principal scale along the meridians are said to be equidistant (French: *projections equidistantes*; German: *abstandstreue Kartennetzentwürfe*).

The list of special properties might include many things. Maling (1968) proposed a list of not less than eleven special properties that may be of practical value. A classification scheme based on such special properties is, however, not advisable. Indeed such a list of special properties has clearly no ending and the resulting classification scheme would become more and more complex. Moreover, some special properties are not mutually exclusive and would therefore lead to overlapping classes.

Though the special properties do not provide a good basis for classification they are nevertheless of importance to the cartographer and should preferrably be referred to in the name of the map projection.

2.4 The Transformation of Map Projections

Once a projection system has been defined mathematically it is possible to perform transformations on it, generally with the purpose of reducing distortion in some particular part or sometimes even over the entire area of the map. A distinction is made between continuous and non-continuous transformations. A transformation of a map projection is said to be continuous when the resulting map represents the spherical surface without discontinuities. Such a transformation can, among others, be accomplished by changing the aspect of a map projection. As seen in section 2.2 a change of aspect does not alter the distortion pattern but merely centers it on another part of the earth's surface with the intention to reduce distortion in the region to be mapped. However this is not the only way by which a continuous transformation can be achieved. Other types of transformation include 'modification' and the 'introduction of parameters in the transformation formulas'.

The non-continuous transformations result in interrupted projections where different parts of the world are represented with their own central axis. In the so-called 'condensed' projections irrelevant areas are even excluded from the map (e.g. when a map has to show a terrestrial distribution, the oceans can be left out). Since experimentation with interrupted projections always goes back to the study of the distortion patterns of the non-interrupted parent graticule(s), these projections will not be discussed in this book.

2.4.1 Modification

On page 16 the possibility was mentioned of replacing the line of zero distortion of the cylindrical and conical projections by two lines along which the principal scale is preserved. Similarly the point of zero distortion of

the azimuthal projection can be replaced by a circle of zero distortion. A transformation of this kind is called a modification. Sometimes the term 'modification' is used in a more general way to cover the whole range of transformations that can be applied to a map projection. However, as Maling (1968) emphasizes, it is necessary from a practical point of view to limit the use of the word 'modification' to those transformations that cause a redistribution of the scale distortion throughout the whole map area without destroying the special properties of the original projection and without affecting the nature and location of singular points.

As an example, the principle of 'modification' will now be applied to the normal aspect of the cylindrical equal-area projection. In that case (Lambert's cylindrical equal-area projection (see Directory, p.159)) parallels and meridians form an orthogonal grid and the line of zero distortion corresponds with the equator along which the principal scale is preserved. In every point on the map $ab = 1$ (equivalence). Since all parallels are represented as straight lines, equal in length to the equator, the x-coordinate of this projection is given by

$$x = R\lambda \qquad (66)$$

The equal-area condition $\sigma = 1$ together with equations (14) and (32) lead to

$$\frac{\mathrm{d}x}{\mathrm{d}\lambda} \frac{\mathrm{d}y}{\mathrm{d}\phi} = R^2 \cos\phi \qquad (67)$$

and with the help of equation (66)

$$\mathrm{d}y = R\cos\phi\,\mathrm{d}\phi \qquad (68)$$

which becomes after integration

$$y = R\sin\phi \qquad (69)$$

Equations (66) and (69) define the rectangular coordinates for the cylindrical equal-area projection in its normal aspect. The scale distortions are

$$h = \frac{\mathrm{d}y}{R\,\mathrm{d}\phi} = \cos\phi \qquad (70)$$

$$k = \frac{\mathrm{d}x}{R\cos\phi\,\mathrm{d}\lambda} = \frac{1}{\cos\phi} \qquad (71)$$

The angular distortion pattern is symmetric about the

equator as can be seen in the Directory, p.158.

The modification of a cylindrical projection consists of the replacement of the line of zero distortion by two lines on which the principal scale is preserved. This can be accomplished by choosing two parallels symmetrical about the equator ($\pm\phi_0$) and represent them equal in length to the corresponding parallels on the generating globe. The x-coordinate of the modified projection then becomes

$$x = R \cos \phi_0 \lambda \qquad (72)$$

Since the equal-area property has to be preserved the y-coordinate is again given by integration of equation (67) and making use of equation (72)

$$y = \frac{R}{\cos \phi_0} \sin \phi \qquad (73)$$

The effect of this modification is a redistribution of the scale distortions

$$h = \frac{dy}{R \, d\phi} = \frac{\cos \phi}{\cos \phi_0} \qquad (74)$$

$$k = \frac{dx}{R \cos \phi \, d\lambda} = \frac{\cos \phi_0}{\cos \phi} \qquad (75)$$

and results in a more even distribution of the distortion values over the total area of the map (e.g. Behrmann's cylindrical equal-area projection (see Directory, p.161)).

A similar modification can be applied to azimuthal and conical projections. It will then lead to distortion patterns characterized by one circle or two circular arcs of zero distortion respectively.

Modified projections are often called secant projections because similar distortion patterns result from perspective projections of the globe upon a cylinder, a cone or a plane that intersects the spherical surface. However this term is regrettable as the principle is applicable in a much more general context.

2.4.2 The Introduction of Parameters in the Transformation Formulas

This kind of transformation has a much greater impact on the distortion pattern of the original projection than modification. Through the introduction of additional parameters in the transformation formulas a more general projection system is obtained which includes the parent projection as special case. The flexibility of

the system depends on the number of parameters that were incorporated in the formulas. The technique is very powerful and permits the manipulation of the distortion pattern to a great extent. Moreover, it is possible to create singular points, e.g. a pole line of given length can be generated from a projection which shows the pole as a point. The transformation can also be applied under certain constraints, e.g. without altering the pattern of areal distortion.

The general theory of this kind of transformation is known in the German literature as *Das Umbeziffern*, see for instance Wagner (1962), Hoschek (1969). McBryde and Thomas (1949) used similar techniques to generalize Eckert's principle for the introduction of so-called 'flat-polar projections'.

To illustrate the concept and its flexibility it will now be applied to Sanson's pseudocylindrical equal-area projection with sinusoidal meridians (Fig. 2.7a). The transformation formulas of this projection are given by

$$\begin{aligned} x &= R\lambda \cos \phi \\ y &= R\phi \end{aligned} \qquad (76)$$

Note that on this equivalent projection the pole is represented as a point while the parallels are correct in length, rectilinear and equally spaced along the central meridian. Suppose now that one wants to represent the entire world within a zone bounded by latitudes $\phi = \pm 60°$ and longitudes $\lambda = \pm 150°$. This can easily be accomplished by multiplying the latitude in the formulas (76) by a factor $m = 2/3$ and the longitude by a factor $n = 5/6$. The transformation formulas of the new graticule become

$$\begin{aligned} x &= R(n\lambda) \cos(m\phi) \\ y &= R(m\phi) \end{aligned} \qquad (77)$$

The transformation preserves the equal spacing of the parallels which characterizes Sanson's original graticule. At the same time, however, the important special property of equivalence is lost. As can be seen from Fig. 2.7b the new graticule, drawn in red, is substantially reduced with respect to the original map. By dividing the x- and y-coordinates by the geometric mean $\sqrt{mn}$ of the parameters which were introduced, the area of the rectangle that encloses the graticule is preserved and the original scale is in some way restored (Fig. 2.7c).

The transformation formulas of the new graticule can now be written as

$$x = R \frac{n\lambda}{\sqrt{mn}} \cos(m\phi)$$

$$(78)$$

$$y = R \frac{m\phi}{\sqrt{mn}}$$

The choice of the parameters m and n determine the curvature of the meridians and the length of the pole line. By varying the value of these parameters an unlimited number of different projections can be generated from the same parent graticule. The original projection is obtained by putting $m = n = 1$ in the general formulas

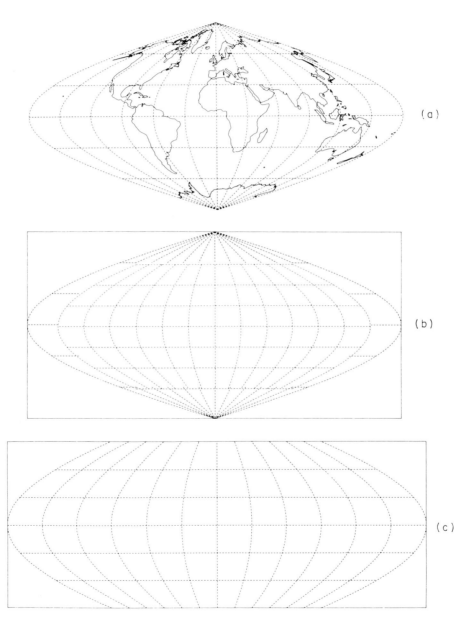

(a)

(b)

(c)

Fig. 2.7 Wagner's general transformation principle applied to Sanson's pseudocylindrical equal-area projection with sinusoidal meridians: (a) Sanson's parent projection; (b) transformation of selected part (shown in red) of Sanson's graticule with preservation of equal parallel spacing; (c) enlargement of the selected part to the original scale of the parent projection.

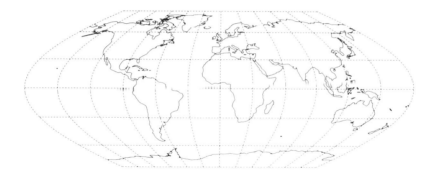

Fig. 2.8   Transformation with equal parallel spacing of Sanson's pseudocylindrical projection.

(78). The transformed map is shown in Fig. 2.8 where it can be seen that the creation of a pole line reduces the angular distortion while the E–W stretching—typical for cylindrical projections—is kept within acceptable limits.

The principle shown in the example above can be generalized to incorporate a wider range of transformations. Let the original graticule be defined as

$$x = f(U, V)$$
$$y = g(U, V) \qquad (79)$$

where U and V correspond with the original geographical coordinates on the globe. Hence

$$U = \phi$$
$$V = \lambda \qquad (80)$$

In the new graticule the transformation formulas can be written as (Wagner, 1962)

$$x = A f(U, V)$$
$$y = B g(U, V) \qquad (81)$$

where U and V are now some function of ϕ and λ

$$U = U(\phi)$$
$$V = V(\lambda) \qquad (82)$$

In principle the functions $U(\phi)$ and $V(\lambda)$ can be chosen freely. However, if certain characteristics of the original graticule are to be maintained, additional constraints have to be imposed upon the choice of these functions. For instance if one wants a proportional scaling along the x- and y-axis between the transformed and the parent projection, as in the example above,

$$\frac{dU}{d\phi} = m \quad \text{and} \quad \frac{dV}{d\lambda} = n \qquad (83)$$

which becomes after integration

$$m = \frac{U}{\phi} \quad \text{and} \quad n = \frac{V}{\lambda} \qquad (84)$$

It is also possible to transform the original graticule without altering the pattern of areal distortion. This type of transformation is very interesting because it allows us to derive new equivalent graticules from existing ones. It will now be shown how the functions (82) have to be adapted to guarantee that the pattern of areal distortion remains unchanged.

According to equations (14), (32) and (79) the areal distortion of the original graticule is given by

$$\sigma_{UV} = \frac{(\partial x / \partial V)(\partial y / \partial U) - (\partial x / \partial U)(\partial y / \partial V)}{R^2 \cos U} \qquad (85)$$

According to equations (14), (32) and (85) the areal distortion of the new graticule becomes

$$\sigma_{\phi\lambda} = \frac{AB[(\partial x / \partial V)(\partial y / \partial U) - (\partial x / \partial U)(\partial y / \partial V)](\partial V / \partial \lambda)(\partial U / \partial \phi)}{R^2 \cos \phi} \qquad (86)$$

If the pattern of areal distortion is not to be altered by the transformation then $\sigma_{UV} = \sigma_{\phi\lambda}$ in every point on the map. Hence

$$\frac{dV}{d\lambda} = \frac{\cos \phi \, d\phi}{AB \cos U \, dU} \qquad (87)$$

Since the right-hand side of equation (87) is independent of λ, it follows that

$$\frac{dV}{d\lambda} = n \qquad (88)$$

and after integration

$$V = n\lambda \qquad (89)$$

Combining equations (87) and (88) and integrating gives

$$\sin U = m \sin \phi \qquad (90)$$

with

$$m = \frac{1}{nAB} \qquad (91)$$

Hence it follows that

$$m = \frac{\sin U}{\sin \phi} \qquad (92)$$

$$n = \frac{V}{\lambda} \qquad (93)$$

$$AB = \frac{1}{mn} \qquad (94)$$

Together equations (92), (93) and (94) define a transformation that maintains the areal distortion pattern of the original projection. Generally the scale factors are distributed evenly, hence

$$A = B = \frac{1}{\sqrt{mn}} \qquad (95)$$

Applied to Sanson's projection the transformation formulas are explicitly written as

$$x = R\frac{n\lambda}{\sqrt{mn}}\sqrt{1 - m^2 \sin^2 \phi}$$
$$\qquad (96)$$
$$y = R\frac{1}{\sqrt{mn}}\arcsin(m \sin \phi)$$

Representing the entire world within a zone of the original graticule bounded by $U = \pm 60°$ and $V = \pm 150°$ leads to the following values of the parameters

$$m = \frac{\sin 60°}{\sin 90°} = \frac{\sqrt{3}}{2} = 0.866$$

$$n = \frac{150°}{180°} = \frac{5}{6} = 0.833$$

$$A = B = 1.177$$

Figure 2.9 shows the new graticule. It is seen that the introduction of the pole line stretches the equatorial areas in the N–S direction and compresses the polar areas in the same direction. It is this phenomenon that makes the cylindrical equal-area projection less suited for world maps. It can only be avoided by violating the equal-area property. Wagner (1962) proposed therefore the following transformation:

$$\sin U = m_1 \sin(m_2\phi) \qquad (97)$$

The application of this transformation to an equal-area projection results in a system with adjustable areal distortion

$$\sigma = \frac{\cos(m_2\phi)}{\cos \phi} \qquad (98)$$

This areal distortion depends on latitude only. Optimizing equation (98) one can demand that the areal

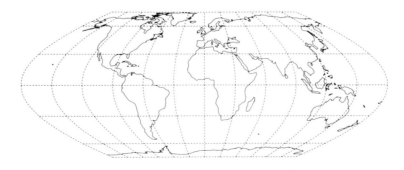

Fig. 2.9 Equal-area transformation of Sanson's pseudocylindrical projection.

distortion does not exceed a given value σ_1 on a given parallel ϕ_1. From equation (98) it then follows that

$$m_2 = \frac{\arccos(\sigma_1 \cos \phi_1)}{\phi_1} \qquad (99)$$

In this way Wagner developed pseudocylindrical projections with pole line and prespecified distortion of areas : $\sigma_1 = 1.2$ for $\phi_1 = 60°$. He suggested that such projections would give a good representation of the populated part of the world (less than 20 per cent of areal distortion). They are intermediate between equal-area projections and pseudocylindrical projections with equally spaced parallels.

Application of this kind of transformation to Sanson's projection leads to the following transformation formulas (Wagner, 1962):

$$x = R\frac{n\lambda}{\sqrt{m_1 m_2 n}} \sqrt{1 - m_1^2 \sin^2(m_2\phi)}$$

$$\qquad (100)$$

$$y = R\frac{1}{\sqrt{m_1 m_2 n}} \arcsin[m_1 \sin(m_2\phi)]$$

Note that the division of the x- and y-coordinates by the factor $\sqrt{m_1 m_2 n}$ does not restore the area of the rectangle which encloses the graticule as before.

Representing the world within the same zone as in the previous example but prespecifying an areal distortion $\sigma = 1.5$ on the parallel of $70°$ leads to the following transformation parameters

$$m_1 = 0.892$$
$$m_2 = 0.845$$
$$n = 0.833$$

Figure 2.10 shows the obtained graticule.

2.5 The Classification of Map Projections

In Section 2.3 some of the disadvantages of a classification scheme, based on special properties, have been enumerated. Clearly a practical scheme should meet the following requirements.

1. The number of classes should be restricted while at the same time the system must show enough flexibility to include an infinite number of possible projections;
2. Overlapping classes should be limited;
3. The classification scheme should facilitate the selection of a projection for a specific purpose.

Tobler (1962) proposed a 'parametric classification' consisting of four groups A–D corresponding to the following mathematical relationships between the coordinates in the plane (u, v) and the geographical coordinates (ϕ, λ):

A	B	C	D
$u = f(\phi, \lambda)$	$u = f(\lambda)$	$u = f(\phi, \lambda)$	$u = f(\lambda)$
$v = g(\phi, \lambda)$	$v = g(\phi, \lambda)$	$v = g(\phi)$	$v = g(\phi)$

Since different aspects of the same projection are not to be considered as different projections (Section 2.2.), only the normal aspect of a projection is used as a criterion for the grouping.

Distinguishing between rectangular coordinates (x, y) and polar coordinates (r, θ) the four groups can be extended to eight functional relationships of which the graphical appearance is illustrated in Fig. 2.11.

An important criterion for the selection of world maps is the appearance of the graticule. Since Tobler's parametric classification has a geometrical meaning it is well suited to underly a practical classification system.

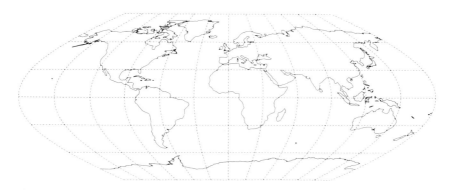

Fig. 2.10   Transformation with adjustable areal distortion of Sanson's pseudocylindrical projection.

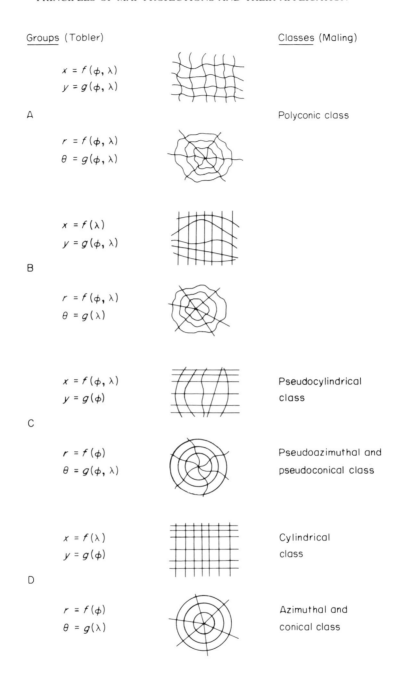

Groups (Tobler)

$x = f(\phi, \lambda)$
$y = g(\phi, \lambda)$

A

$r = f(\phi, \lambda)$
$\theta = g(\phi, \lambda)$

$x = f(\lambda)$
$y = g(\phi, \lambda)$

B

$r = f(\phi, \lambda)$
$\theta = g(\lambda)$

$x = f(\phi, \lambda)$
$y = g(\phi)$

C

$r = f(\phi)$
$\theta = g(\phi, \lambda)$

$x = f(\lambda)$
$y = g(\phi)$

D

$r = f(\phi)$
$\theta = g(\lambda)$

Classes (Maling)

Polyconic class

Pseudocylindrical class

Pseudoazimuthal and pseudoconical class

Cylindrical class

Azimuthal and conical class

Fig. 2.11   Tobler's parametric classification scheme with corresponding classes according to Maling (after Tobler, 1962, and Maling, 1973).

Nevertheless groups C and D comprise projections that show serious differences in appearance even though they satisfy the same functional relationships. Maling (1973) therefore subdivides the map projections in seven different classes (Fig. 2.12) that can be related to Tobler's major subdivision (Fig. 2.11):

1. The polyconic class: comprises all the map projections of group A. Both parallels and meridians of the grid are curved.
2. The pseudocylindrical class: contains the projections which have curved meridians and straight parallels. The class consists of the projections of group C

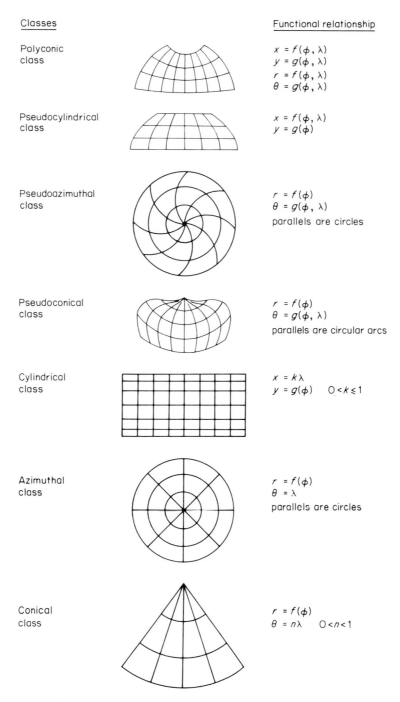

Fig. 2.12   The seven classes of Maling's classification.

defined in terms of Cartesian coordinates. They are especially important for world maps.

3. The pseudoazimuthal class: contains the projections which have curved meridians and parallels consisting of concentric circles. It is the only class that contains no projections of practical significance. From a theoretical point of view, however, the projection of Wiechel might be mentioned since it combines the equal-area property with equidistance along the meridians (Arden-Close, 1952).

4. The pseudoconical class: represents meridians as concurrent curves and parallels as concentric circular arcs. The angle θ between the meridians at the center of the projection is always smaller than the corresponding spherical angle λ so that the parallels cannot form a complete circumference. Both the pseudoconical and the pseudoazimuthal class are defined in terms of polar coordinates and form part of group C.

5. The cylindrical class: comprises projections where both parallels and meridians are represented by parallel straight lines. These orthogonal grids are defined in terms of Cartesian coordinates and form part of group D.

6. The azimuthal class: represents meridians as concurrent straight lines and parallels as concentric circles. The center of the parallels coincides with the point of concurrency of the meridians. All angles measured at the center of the projection are correctly represented.

7. The conical class: differs from the previous one in the representation of the azimuthal angles. For conical projections the angle θ on the map is always smaller than the corresponding angle λ on the sphere. The parallels are therefore concentric arcs and the characteristic outline of the projection is fan-shaped. Together with the azimuthal projections the conical projections belong to those projections of group D which are defined in terms of polar coordinates.

Note that no projections of group B were included in the system of Maling. This group contains a few projections for specific geographic purposes that are defined in terms of polar coordinates (Tobler, 1968). In conventional cartography, however, these projections are seldom used. It is obvious that what is mentioned about the nature of meridians and parallels also applies to the transformed meridians and parallels in the transverse and oblique aspects of a projection.

The seven above-mentioned classes provide a good basis for the selection of world maps. Each class corresponds to a collection of projections with similar appearance and distortion characteristics. Hence, the choice of the class can be considered as the first step in the selection of a world map. Maling makes a further threefold subdivision in each class according to a constant, decreasing or increasing separation of the parallels for equal increments of latitude. This principle, which holds for the parallel spacing in the normal aspect of a projection, also holds for the spacing of the transformed parallels in the transverse and oblique aspects. Since many special properties (equivalence, conformality) are related to some extent to the spacing of the parallels the subdivision has a practical benefit. Nearly all equal-area projections show a decreasing parallel spacing with increasing latitude. The conformal projections, on the other hand, are characterized by an increased spacing of the parallels. With respect to their distortion characteristics, projections with equally spaced parallels generally occupy an intermediate position within the continuum of all map projections where the equal-area and conformal projections can be considered as the limiting cases.

Maling's seven times three subdivision will be adopted in this book, although it must be realized that the classification includes some pitfalls. For example, Sanson's equal-area projection is pseudocylindrical and at the same time a limiting case of Bonne's pseudoconical projection with the standard parallel coinciding with the equator. Table 2.1 gives the classification of all projections dealt with in this book. Note that only four of Maling's seven classes were actually used, since the other three (the azimuthal, pseudoazimuthal and conical class) are not suited for general-purpose world maps. Also an additional distinction was made between projections with and without a pole line. This subdivision was added with regard to the map selection process.

2.6 The Nomenclature of Map Projections

In the nomenclature, associated with the classification system, the name of each projection should reflect the appearance of the graticule. The naming should also be brief, easy to comprehend and unequivocal.

In the past projections were often named after their 'inventor'. This has led to confusion since there is not seldom disagreement on who is the original author of a projection. Moreover, several projections may be due to the same person. Very often it then concerns projections with minor differences in appearance. At best those projections are numbered, e.g. Eckert I to Eckert VI. Although such a numbering helps identification it appeals only to the memory. Numbering has clearly no relation to the appearance of the grid system or the special properties involved.

Maling (1968) proposed a series of descriptive terms that should lead to an unequivocal identification of a projection:

Table 2.1 Overview of all projection systems discussed in the Directory

	Polyconic projections		Pseudocylindrical projections		Pseudoconical projections	Cylindrical projections
	Pole = point	Pole = line	Pole = point	Pole = line	Pole = point	Pole = line
Equally spaced parallels	Aitoff	Aitoff-Wagner Winkel–Tripel	Apianus II/ Arago Putnins P1 Putnins P3 Sanson	Ortelius Eckert III Wagner VI/Putnins P1' Kavraisky VII Winkel II Eckert IV Wagner III Winkel I Eckert I	Bonne Werner (cordiform projection)	Cylindrical equidistant projection: Plate *carée* Equi-rectangular projection
Decreasing parallel spacing	Hammer–Aitoff/ Nordic/Briesemeister Eckert-Greifendorff	Hammer–Wagner	Mollweide/Atlantis Putnins P2 Boggs Craster = Putnins P4 Adams Kavraisky V	Eckert IV Wagner IV = Putnins P2'/Werenskiold III Wagner V Putnins P4'/Werenskiold I Flat-polar parabolic authalic Flat-polar quartic authalic Flat-polar sinusoidal authalic Eckert VI Wagner I = Kavraisky VI/Werenskiold II Wagner II Nell–Hammer Eckert II Robinson		Cylindrical equal-area projection: Lambert Behmann Peters Pavlov
Increasing parallel spacing	Lambert van der Grinten I			Ginsburg VIII		Mercator Miller I/Miller II Cylindrical stereographic projection: Braun BSAM Gall Urmaev III

1. The projection class: this allows to locate the grid within the classification scheme and, by the same token recognition of the distortion patterns.
2. The aspect: reference to the aspect is necessary for all transverse and oblique variations of a projection, because the aspect affects the appearance of the graticule. In addition it might be necessary to provide further information on the projection center, the orientation of the axes of the distortion pattern, etc.
3. Special properties: they are relevant in relation to the potential use of the projection system.
4. Distinctive characteristics: they provide additional information which should make a complete identification possible, such as the nature of modification, the characteristic outline of the map, the general appearance of meridians, parallels or poles, etc.

In some cases, however, a lengthy description still fails to define a projection without ambiguity. Therefore it might be necessary to combine the descriptive terms with a reference to the author. This is also the case for those projections which are characterized by a long history. In this respect it is important not to disorientate the reader who has some knowledge of map projections and who is accustomed to the use of certain names (e.g. Mercator's projection). The same applies for special references given by certain authors to identify their own inventions (e.g. flat polar quartic authalic projection, Bartholomew's Nordic and Atlantis projection).

3

THE EVALUATION AND CHOICE
OF
A MAP PROJECTION

The selection of a suitable map projection is one of the many variables involved in cartographic design and has to be integrated with other decisions to be made such as the content of the map, the type of symbols to be used, the scale, the potential use, etc. (Mei-Ling Hsu, 1981).

The importance of a proper choice of projection increases as the scale of the map decreases. This is especially so for world maps where the requirements might vary considerably from one application to another. Although the number of existing projections is enormous, very often none of them satisfies a given set of constraints. In that case it might even prove necessary to develop a new projection or to modify an existing one.

The evaluation of a map projection generally consists of a quantitative and a qualitative analysis.

3.1 Quantitative Analysis

The quantitative analysis concerns the evaluation of map projections according to their distortion characteristics. The choice of criteria to accomplish this evaluation depends on the intended purpose of the map.

Generally there are two possibilities:

1. The map has to be equivalent or conformal. Within the whole range of map projections, conformal and equal-area projections may be considered as two extremes. Conformality and equivalence are mutually exclusive : a conformal projection normally shows a considerable distortion of areas while an equal-area projection is similarly characterized by a large angular distortion.

2. When none of these special properties are required it is possible to select projections that have intermediate distortion characteristics. Generally such projections have a smaller overall linear scale distortion than the conformal and equal-area projections.

The demand for an equal-area or a conformal projection always leads to extreme distortions near the edges of the map. This is particularly true for world maps, and as these maps are seldom used for precise measurements conformality and equivalence are more and more abandoned in this field of cartography.

3.1.1 Criteria for the Evaluation of Map Projections

The deformation of a map projection can be regarded as being the result of the linear scale distortion, which varies with direction and from point to point. Hence, in order to evaluate map projections, it seems appropriate to integrate the distortion over the whole map area. In Section 1.3 it was shown how the full range of deformational characteristics in a certain point on the map could be portrayed by means of Tissot's indicatrix, an ellipse resulting from the projection of an elementary circle centered round the corresponding point on the generating globe. As said this ellipse is characterized by a semimajor axis a and semiminor axis b, corresponding with maximum and minimum linear distortion respectively.

1. The areal distortion is defined by the ratio of the area of the indicatrix to the area of the elementary circle and is given by $\sigma = ab$ (equation 53). If no areal distortion occurs $\sigma = 1$. A criterion for the areal distortion over a well determined part S of the map is given by

$$\int_S \int (ab - 1)^2 \mathrm{d}S \qquad (101)$$

For an equal-area map this criterion is zero.

2. The angular distortion is more difficult to quantify since varies with direction. As shown on Fig. 1.8, it reaches a maximum for the angles, subtended by the directions of maximum distortion. Integrating expression (54) for the maximum angular distortion over the mapping area S gives

$$\int_S \int \left(2\arcsin\frac{a - b}{a + b}\right) \mathrm{d}S \qquad (102)$$

This criterion is zero for conformal projections, where $a = b$ for every point on the map.

3. A similar difficulty arises with the linear distortion or the scale factor. As said above, this value varies with direction from a minimum b to a maximum a. For $a = 1$ or $b = 1$ there is no linear distortion along the corresponding principal direction. A frequently used index to quantify the scale factor in a point is given by

$$(a - 1)^2 + (b - 1)^2 \qquad (103)$$

which is the sum of the squares of the scale errors along the principal directions. A criterion for the scale distortion over the mapped area is then given by

$$\int_S \int \left[(a - 1)^2 + (b - 1)^2\right] \mathrm{d}S \qquad (104)$$

This expression always differs from zero.

The squares in expressions (101) and (104) assure that all terms in the integration are positive. This is a necessary condition to prevent enlargements and reductions from compensating each other.

Airy made already use of this concept of integration in 1861 (Maling, 1973; Snyder, 1985). He applied the least-squares technique (cf. equation 104) to obtain the so-called azimuthal projection with a minimum 'total misrepresentation' by 'balance of errors' (Section 3.1.2). Behrmann (1909), on the other hand, applied equation (102) in an effort to compare a number of equal-area projections by calculating the global mean of the maximum angular distortion. Table 3.1 shows Behrmann's results. He concluded that the cylindrical equal-area projection with two standard parallels at $\pm 30°$ latitude was the best choice for a global representation of the earth.

The use of mean distortion values can be further improved through the concept of 'areas of major interest' (Robinson, 1951). Robinson stated that the determination

Table 3.1 Mean maximum angular distortion from some equal-area projections (after Behrmann, 1909).

Behrmann's results (1909)		Mean maximum angular distortion
Cylindrical equal-area projections		
Standard latitude:	0°	31°25′
	10°	29°51′
	20°	28°05′
	30°	27°06′
	40°	29°46′
	50°	38°40′
	60°	55°01′
Other equal-area projections		
Eckert's pseudocylindrical equal-area projection with elliptical meridians (Eckert IV)		27°34′
Mollweide's pseudocylindrical equal-area projection		32°07′
Eckert's pseudocylindrical equal-area projection with sinusoidal meridians (Eckert VI)		32°19′
Hammer's polyconic equal-area projection (Hammer–Aitoff projection)		37°34′
Eckert's pseudocylindrical equal-area projection with rectilinear meridians (Eckert II)		38°18′
Sanson's pseudocylindrical equal-area projection		38°40′
Lambert's azimuthal equal-area projection		49°40′
Collignon's pseudocylindrical equal-area projection		55°23′

of a mean distortion value over the entire surface of the map is not very realistic when the region of interest covers only part of it. This is the case for most thematic world maps, where usually the area of major interest covers only the continents.

3.1.2 Minimum-error Projections

The same quantitative evaluation criteria can also be applied to study the impact of transformations (Section 2.4) on the distortion characteristics of a map projection. For instance, the 'optimal aspect' of a projection can be selected by minimizing the integrated value of a distortion parameter over the area of interest. The modification of a projection through the introduction of two lines of zero distortion (in the case of cylindrical and conical projections) or one standard circle (azimuthal projections) can be optimized in a similar way. However, the most general and interesting use of this minimization technique occurs when introducing a number of parameters in the transformation formulas of a projection and optimizing their values under certain constraints (e.g. a special property, the length of a pole line, the ratio of the axes, etc.).

The classical example of a minimum-error projection is Airy's azimuthal projection mentioned earlier (Section 3.1.1). Most papers on minimum-error projections were published between 1850 and 1950 (Snyder, 1985). However, they were restricted with respect to the number of parameters to be optimized (generally restricted to one) and to those map projections which could easily be defined in mathematical terms. The full possibilities of the least-squares technique in map projection research could only be exploited with the advent of high-capacity computers. Recently Snyder (1985) used the technique with complex algebra to create a low error conformal projection for a 50-state map of the United States. Apart from Airy's criterion, many other criteria have been used for the optimization process. Kavraisky (Grafarend and Niermann, 1984; Snyder, 1985) for instance used a modification of Airy's least-squares criterion

$$\int_s \int \left[(\ln a)^2 + (\ln b)^2 \right] \mathrm{d}S \qquad (105)$$

where the logarithm assures that enlargements and reductions by the same factor have the same weight in the summation.

Although the term 'minimum-error' is generally restricted to projections that result from application of the least squares principle, it is obvious that the concept can be extended to other criteria. For example, Behrmann's cylindrical equal-area projection with standard parallels at 30° N and S (Behrmann, 1909) results from optimizing the standard latitude of the cylindrical equal-area projection by means of equation (102) (Table 3.1).

3.1.3 Limitations of the Quantitative Analysis

The quantitative analysis proves to be a very reliable method for the selection of a large-scale map projection. However, as the scale of the map decreases other considerations, especially those related to geographical perception, become more and more important. Behrmann's comparative analysis of equal-area projections illustrates the conflict that rises between the geographical and the abstract-mathematical approaches. He concluded that the cylindrical equal-area projection with two standard parallels at 30° N and S is the 'best' available projection for world maps, because it has the lowest mean angular distortion of all projections considered. A brief look at this map (see Directory, p.160), however, shows the extreme E–W stretching of the higher latitudes. Moreover, and this applies for all cylindrical projections, there is a complete lack of any suggestion that the earth has a spherical shape.

From a perceptive point of view, a world map should give an image of the earth that closely resembles the configuration and the shape of the continents on the globe. This implies that the continents should not be interrupted and that the deformation of large shapes should be kept as low as possible. However, the problem is that the deformation of large shapes is difficult to quantify. Tissot's theory of distortion is a local theory and a projection with a low angular distortion only gives a good representation of shape for a small area around every point on the map. Thus whereas Tissot's theory provides us with the elegance of a straightforward mathematical approach, many arguments can be raised against it when it comes to the evaluation of the representation of areas with continental scale.

More recently, several attempts have been made to quantify finite distortions. Tobler (1977) developed a minimum-error projection for the United States by minimizing the mean linear distortion of all distances between the points of a regular grid, covering the area of interest (Snyder, 1985). Peters (1975) used a similar approach to optimize the application of Wagner's transformation principle (Section 2.4) to existing map projections. However, instead of working with a grid, he minimized the mean linear distortion of 30 000 random distances covering the entire surface of the earth. Later he refined his technique by considering only those distances connecting two points on the continental surface (Peters, 1978). The results of the Peters' analysis are shown in Table 3.2. His so-called 'Entfernungsbezogene Weltkarte', which has the lowest mean linear distortion of all equal-area projections considered, is obtained through optimization of the parameters of Wagner's polyconic equal-area projection with pole line (Hammer–Wagner projection (see Directory, p.67)). Although the Peters analysis is very

interesting, his method still does not solve the problem of the deformation of large continental shapes. Peters concludes that giving up the equal-area property can only lead to a slight improvement of the mean linear distortion and therefore dedicates his further attention to equal-area projections only. A visual comparison of the maps involved in the analysis, however, shows that non-equal-area projections give a better portrayal of large shapes than equal-area projections with similar distortion values.

From the foregoing it is clear that any attempt to develop adequate world maps by minimum-error techniques can only be succesful if certain perceptive considerations are taken into account during the minimization process. Therefore a qualitative analysis has to be carried out first of all in which the perceptive (and other) requirements are identified and translated into a set of operational constraints.

Table 3.2 Mean linear distortion for selected world map projections (after Peters, 1978).

Peters' results (1978)	Mean linear distortion(%)
Equal-area projections	
'Entfernungsbezogene Weltkarte'	11.0
Wagner's polyconic equal-area projection with pole line (Hammer–Wagner projection)	11.5
Mollweide's pseudocylindrical equal-area projection	12.0
Hammer's polyconic equal-area projection (Hammer–Aitoff projection)	12.2
Sanson's pseudocylindrical equal-area projection	13.7
Behrmann's cylindrical equal-area projection with standard latitude $30°$	14.3
Other projections	
Winkel's polyconic projection with equally spaced parallels and pole line (Winkel–Tripel projection)	10.9
Aitoff's polyconic projection with equally spaced parallels	11.7
Kavraisky's pseudocylindrical projection with equally spaced parallels, elliptical meridians and pole line (Kavraisky VII)	11.8
Cylindrical equidistant projection with standard latitude $30°$	14.2
Mercator's cylindrical conformal projection	29.1

3.2 Qualitative Analysis

The qualitative analysis involved in the selection of a suitable projection for a world map deals with items as the choice of a special property, the representation of large shapes and problems of design. The finally selected projection results from a careful balancing of these closely interrelated factors while taking into account the results of the quantitative analysis.

3.2.1 The Choice of a Special Property

In topographic mapping the use of conformal projections is universally accepted. In small-scale cartography (e.g. atlas cartography) the special property of equivalence has for a long time been regarded as the most important property, especially when it concerns the mapping of statistical data. More recently, however, world map projections, which are neither conformal nor equivalent but occupy an intermediate position between these two extremes, are becoming increasingly popular. Their associated favorable distortion patterns result from balancing the angular and areal distortions and appear especially conspicuous near the edges of the map. This implies that these projections give a far better portrayal of large shapes. Equal-area projections are henceforth recommended only in the cases where areal measurements are strictly needed or where density data are represented by point symbols.

3.2.2 *The Representation of Large Shapes*

It has already been said that distortion theory cannot deal with the deformation of large shapes like continents or oceans and that the representation of these shapes is strongly related to the choice of a special property.

When a conventional arrangement of meridians and parallels is not required an oblique aspect can cause a great improvement by placing the large configurations that are of interest in the least distorted parts of the map. For example, when the map aims at representing terrestrial distributions (soils, vegetation, etc.) the location of the origin and the orientation of the axes can be chosen in such a way that the distortions are minimized over the continental area.

3.2.3 *Problems of Design*

While the choice of special property and aspect can be achieved in a more or less rational way there always remain some decisions to be made which cannot be translated into specific rules. Such decisions are strongly related to the experience and 'good taste' of the designer and interfere with the rational decisions. For example the use of an oblique aspect of a projection may lead to a better representation of the continents, but it is the author's decision to take such an option which results in a more unusual representation. Generally the normal aspect is preferred because of the geometrically simpler graticule. Indeed, deciding on the nature of meridians and parallels is not always self-evident. A cylindrical map might be suggested for multipurpose maps as the grid system is excellent for referencing. However, a rectangular grid lacks any suggestion of the spherical shape of the earth and a projection with curved meridians might therefore be preferred. Pseudocylindrical projections provide then a good alternative but the diversity in this class of projections is enormous. Here the mapmaker will have to decide on the exact nature of the meridians, the spacing of the parallels and if the pole is to be shown as a point or as a line. The latter decision is especially important due to the excessive compression of the polar areas on projections that represent the pole as a point. Indeed, this can be remedied by the introduction of a pole line, but has the disadvantage of creating a discontinuity of the meridians at the pole. Once again it is the designer of the map who has to take a decision on this matter.

The main reason why pseudocylindrical projections are so frequently used for world maps is that they combine curved meridians, which suggest the sphericity of the earth, and straight-line parallels, which allow easy latitudinal referencing. However, the curvature of both meridians and parallels simulates better the spherical earth and leads to more favourable distortion patterns.

Finally, the overall shape of the projection must—in many cases—suit a given format. This influences projection characteristics such as the nature of the meridians and especially the ratio of the axes.

3.3 The Choice of a Map Projection in Practice

A map can be considered as a powerful means of communication by which space related information is conveyed. As one of the decisions to be made in cartographic design, the choice of projection should therefore be strongly related to the final purpose and use of the map. Generally the cartographer which has to decide on the choice of a map projection has two ways to follow:

1. He analyses the problem and comes to a unique set of requirements. Within this set of constraints he designs a new projection that suits best the purpose in mind. Clearly, this is the best option but in practice one can say that only a few projections have been developed in this way. Indeed, the design of a new projection demands considerable skill and a non-negligable effort from the cartographer.
2. He analyses the problem, comes to a set of requirements and tries to select a projection within the given constraints. Although this is the way usually followed in cartographic design it does not always lead to an ideal solution. In many cases, however, the compromise will be satisfactory and will not justify the effort needed to design a new projection which very often only comprises minor improvements.

Whatever the option taken the procedure followed for the selection of a map projection involves succesively a qualitative and a quantitative analysis.

During the qualitative analysis decisions are made on the following points: special property, class of projection, the pole = point or pole = line dilemma, the aspect (normal, transverse, oblique), etc. In this way the number of possible projections is drastically reduced and the further quantitative analysis might then help to select the most suitable projection within the given constraints. The quantitative analysis is the appropriate method to decide on the origin and orientation of the axes in case of an oblique aspect. It also helps to evaluate the impact of a modification or of the introduction of a pole line on the distortion pattern.

At present the role of the quantitative approach should not be overrated. The developed evalutation techniques are useful, but the ultimate choice still highly depends on the skill and experience of the cartographer. A lot of factors that are important in the selection process are not yet susceptible to quantitative measurement. However, the growing importance of fully automated cartography and geographical information systems calls for a more objective selection process. Therefore more research on the formalization of the human decision processes involved in map design is greatly needed.

Part II

DIRECTORY
OF
WORLD MAP PROJECTIONS

4

INTRODUCTION

The second part of this book gives an overview of 68 projections representing the earth as a whole. The main purpose is to provide for each projection ready-to-calculate formulas as well as the necessary information to evaluate and to situate the projection within the complex range of possible map projections. This directory is certainly not complete. The design of new map projections has always been a great challenge and the number of existing world graticules is therefore enormous. However, few graticules were designed with a specific purpose and many of them were never implemented in the cartographic practice. The projections, retained in this directory, were chosen either because of their practical value (e.g. the Winkel–Tripel projection), their general familiarity (e.g. Sanson's pseudocylindrical equal-area projection) or their historical importance (e.g. the cordiform projection). Some projections have little significance of their own and were merely included because they underlie the construction of other projections which are better suited for world maps (e.g. the plate *carrée*).

The information given with each projection in the directory includes the following items :

1. The name of the projection in descriptive terms. This allows us to situate the projection in the classification table (Part I, Table 2.1).
2. Possible alternative names which are commonly used.
3. The name of the author and—if possible—reference to the original paper.
4. A short description of the visual appearance, distortion characteristics and history of the projection.
5. The transformation formulas.

6. The mean distortion values D_{ar}, D_{arc}, D_{an}, D_{anc}, D_{ab}, D_{abc}.
7. One or two examples of a world map showing the continental configuration with superposition of lines of equal areal and/or equal angular distortion.

Some projections are shown in different aspects, others are modified. These examples illustrate the great flexibility that exists among the different ways by which the earth can be portrayed.

4.1 Transformation Formulas

The formulas to transform the geographical coordinates ϕ, λ into Cartesian x-, y-coordinates are generally kept as simple as possible. The origin of the cartesian coordinates is always the centre of the map. Latitude ϕ and longitude λ are expressed in radians. Latitude is reckoned positive in the Northern Hemisphere and negative in the Southern Hemisphere. Longitude is reckoned from the central meridian which passes through the mapping centre, counted positive to the east, negative to the west. Although the central meridian can be any meridian, in this directory it is generally chosen to be the Greenwich meridian. Some examples are given where the central meridian has been shifted in order to illustrate a less 'classical' view of the earth.

The dimensions of the map to be produced depend on the scale and can be manipulated in the formulas by means of R, which is the radius of the generating globe. As explained in Part I (Section 1.1.) the nominal scale S_N is the radius of the generating globe to the radius of the earth (R_E = 6371 km). Thus, if a map is required at a scale of 1 : 90 000 000

$$R = \frac{6371.10^5 \text{ cm}}{90.10^6} = 7.1 \text{ cm}$$

The dimensions of the map are then easily found from the transformation formulas: for instance, if the cylindrical stereographic projection with two standard parallels ($\phi_0 = \pm45°$) is required the formulas (p.169) give

$x = R\pi\, 0.707 = 15.8$ cm   for $\phi = 0$, $\lambda = \pi$

$y = R(1 + 0.707) = 12.1$cm   for $\phi = \pi/2$, $\lambda = 0$

so that the world map is contained in a rectangle with dimensions 31.6 × 24.2 cm.

Inversely, if the format of the map is given from the start, the radius of the generating globe and the scale can be found in a similar way. Suppose, one wants to make a pseudocylindrical map (say Eckert III) on a sheet of paper which has dimensions 30 × 20 cm. Since Eckert III is a projection characterized by a 2 : 1 ratio of the axes, a suitable choice for the radius r of the inner hemisphere might be $r = 6.5$ cm which gives as longer axis of the map 26 cm and as smaller axis 13 cm. The radius R of the generating globe is now calculated as (see formulas, p. 85)

$$R = r\sqrt{\frac{4 + \pi}{4\pi}} = 4.9 \text{ cm}$$

which in turn allows to calculate the nominal scale

$$S_N = \frac{4.9 \text{ cm}}{6371.10^5 \text{ cm}} \approx \frac{1}{130\,000\,000}$$

4.2 Patterns of Distortion

Lines representing equal values of maximum angular distortion and areal distortion were superposed on the basic maps, which show the continental configuration.

Maximum angular distortion 2Ω is defined by (Part I, Section 1.3.2)

$$2\Omega = 2 \arcsin \frac{a - b}{a + b} \tag{54}$$

and varies between 0 and 180°. On conformal projections this value is zero in every point of the map. Since it represents the maximum angular distortion which can occur in a point it is expressed as a function of maximum and minimum linear distortion a and b, which are dimensionless numbers (the linear distortion equals unity when no distortion takes place). The areal distortion is also defined in function of a and b and equals (Part I, Section 1.3.2)

$$\sigma = ab \tag{55}$$

σ varies between zero and infinity, being >1 for areal enlargement and < 1 for areal reduction. For equal-area projections $\sigma = 1$ in every point of the graticule.

4.3 Distortion Parameters

In addition to the distribution of maximum angular deformation and areal deformation each map in the Directory is characterized by six distortion parameters, D_{an}, D_{ar}, D_{ab}, D_{anc}, D_{arc}, D_{abc}.

Three of them are obtained by numerical integration of the angular, the areal and the scale distortion respectively over the total area of the map. They are defined as follows

$$D_{an} = \frac{1}{S} \sum_{j=1}^{m} 2 \arcsin\left(\frac{a_j - b_j}{a_j + b_j}\right) \cos \phi_j \Delta\phi\Delta\lambda \tag{106}$$

$$D_{ar} = \frac{1}{S} \sum_{j=1}^{m} \left[(a_j b_j)^p - 1 \right] \cos \phi_j \Delta\phi\Delta\lambda \tag{107}$$

$$D_{ab} = \frac{1}{S} \sum_{j=1}^{m} \left[\frac{a_j^{\,q} + b_j^{\,r}}{2} - 1 \right] \cos \phi_j \Delta\phi\Delta\lambda \tag{108}$$

where m is the number of grid points and

$$S = \sum_{j=1}^{m} \cos \phi_j \Delta\phi\Delta\lambda \tag{109}$$

is the total area to be mapped.

$$p \begin{cases} = 1 & \text{if } a_j b_j \geq 1 \\ = -1 & \text{if } a_j b_j < 1 \end{cases}$$

$$q \begin{cases} = 1 & \text{if } a_j \geq 1 \\ = -1 & \text{if } a_j < 1 \end{cases}$$

$$r \begin{cases} = 1 & \text{if } b_j \geq 1 \\ = -1 & \text{if } b_j < 1 \end{cases}$$

$\Delta\phi$, $\Delta\lambda$: interval in latitude and longitude between grid points for which the distortions are calculated ($2.5°$ for all computations in this directory).

D_{an} is the mean angular deformation obtained by averaging the maximum angular deformation 2Ω over all grid points and weighted with respect to the elementary area surrounding each point.

D_{ar} is the weighted mean error in areal distortion. This error is defined as the deviation from the distortion-free value $a_j b_j = 1$. The coefficient p ensures that reductions and enlargements with the same amount have equal weight in the summation.

D_{ab} is the weighted mean error in the overall scale distortion. The scale distortion in a certain point is defined as the mean of the maximum and minimum scale distortion a and b and the error as the deviation of this mean from 1. The coefficients q and r in equation (108) have the same function as p in expression (107).

The parameters D_{ar} and D_{ab} are thought to be more realistic for world maps than the generally used criteria (101) and (104). The latter formulas are well suited for maps at larger scale. However, for world maps which are characterized by a wider range of distortion values the squares in equations (101) and (104) overrate the extreme distortions. The parameters D_{ar} and D_{ab} avoid this problem as can be seen in the following example: consider the cylindrical equidistant projection with two standard parallels which is a modification of the well-known 'plate *carrée*' or simple cylindrical projection. Minimization of D_{ab} with respect to the standard latitude leads to an optimal value with $\phi_0 = 37.5°$. Minimization of equation (104), on the other hand, leads to an optimal standard latitude $\phi_0 = 69.4°$. Figure 4.1 shows both results. The positioning of the standard parallel at an extreme high latitude in the second case results from the overrating of the high distortion values and produces a grid that elongates excessively the N–S distances in lower latitudes.

As explained more extensively in Part I (Section 3.1) a thematic map very often deals with terrestrial distributions so that for world maps it might prove advantageous to calculate the distortion values over the continents only. This is accomplished by multiplying the expressions (106), (107) and (108) by a factor P_j which indicates whether a grid point j represents continental surface ($P_j = 1$) or not ($P_j = 0$). The distortion parameters D_{anc}, D_{arc} and D_{abc} (c for continental) are then given by

$$D_{anc} = \frac{1}{S_c} \sum_{j=1}^{m} 2 \arcsin\left(\frac{a_j - b_j}{a_j + b_j}\right) P_j \cos\phi_j \, \Delta\phi\Delta\lambda \tag{110}$$

$$D_{arc} = \frac{1}{S_c} \sum_{j=1}^{m} \left[(a_j b_j)^p - 1\right] P_j \cos\phi_j \, \Delta\phi\Delta\lambda \tag{111}$$

$$D_{abc} = \frac{1}{S_c} \sum_{j=1}^{m} \left[\frac{a_j^q + b_j^r}{2} - 1\right] P_j \cos\phi_j \, \Delta\phi\Delta\lambda \tag{112}$$

with

$$S_c = \sum_{j=1}^{m} P_j \cos\phi_j \Delta\phi\Delta\lambda \tag{113}$$

representing the continental area to be mapped.

The integration of distortion over so-called areas of major interest (Robinson, 1951) allows very often a more realistic evaluation to be made. In addition it becomes possible to experiment with different aspects and eventually to select a certain map projection according to minimum values of the distortion parameters D_{anc}, D_{arc} and D_{abc}. The examples given in this directory show that a change of aspect allows a reduction of the mean distortion values for a world map by a factor of two.

Table 4.1 gives an overview of the six distortion values calculated for each projection described in the directory. These results are graphically represented in Fig. 4.2 where the mean angular distortion is plotted versus the mean areal distortion. Both were obtained after integration over continental area. The figure shows that projections with small angular distortion are generally characterized by a large areal distortion and vice versa. The grouping of points according to the nature of the spacing of the parallels shows how the latter allows a quick evaluation of the distortion characteristics. Conformal projections as well as projections with small angular distortion are characterized by increased parallel spacing (from the equator towards the poles), while

Table 4.1 Overview of distortion parameters for the projections discusssed in the Directory

		Polyconic projections	D_{ar}	D_{arc}	D_{an}	D_{anc}	D_{ab}	D_{abc}
	Equally spaced parallels	Aitoff	0.23	0.19	30.2	28.9	0.36	0.34
Pole = point	Decreasing parallel spacing	Hammer–Aitoff	0.00	0.00	35.7	33.6	0.43	0.41
		Nordic	0.00	0.00	35.7	23.7	0.43	0.26
		Briesemeister	0.00	0.00	35.8	24.0	0.47	0.26
		Eckert-Greifendorff	0.00	0.00	35.5	35.6	0.45	0.46
	Increasing parallel spacing	Lambert	1.38	1.73	0.0	0.0	0.49	0.59
		van der Grinten I	1.87	1.87	7.7	7.3	0.67	0.67
Pole = line	Equally spaced parallels	Aitoff–Wagner	0.33	0.42	21.2	22.0	0.26	0.29
		Winkel–Tripel	0.17	0.25	23.4	22.7	0.25	0.26
	Decreasing parallel spacing	Hammer-Wagner	0.00	0.00	30.7	30.7	0.36	0.38
	Increasing parallel spacing							

Pseudocylindrical projections	D_{ar}	D_{arc}	D_{an}	D_{anc}	D_{ab}	D_{abc}
Apianus II	0.23	0.31	24.2	25.5	0.30	0.33
Arago	0.21	0.25	24.2	25.5	0.31	0.34
Putnins P1	0.10	0.11	30.8	31.4	0.39	0.40
Putnins P3	0.04	0.05	35.1	35.2	0.43	0.45
Sanson	0.00	0.00	39.0	38.0	0.51	0.49
Mollweide	0.00	0.00	32.3	33.7	0.39	0.42
Atlantis	0.00	0.00	32.3	22.0	0.39	0.23
Putnins P2	0.00	0.00	34.9	35.5	0.50	0.55
Boggs	0.00	0.00	35.0	35.1	0.44	0.44
Craster = Putnins P4	0.00	0.00	37.0	36.6	0.48	0.47
Adams	0.00	0.00	36.0	36.4	0.47	0.48
Kavraisky V	0.00	0.00	30.5	32.9	0.38	0.43

Pseudoconical projections	D_{ar}	D_{arc}	D_{an}	D_{anc}	D_{ab}	D_{abc}
Bonne	0.00	0.00	43.2	25.8	0.60	0.32
Werner	0.00	0.00	45.7	27.8	0.64	0.34

	D_{ar}	D_{arc}	D_{an}	D_{anc}	D_{ab}	D_{abc}
Ortelius	0.39	0.54	21.6	25.2	0.29	0.38
Eckert III	0.36	0.43	18.2	20.8	0.28	0.33
Wagner VI	0.33	0.46	20.4	22.2	0.26	0.31
Putnins P1'	0.27	0.37	20.4	22.2	0.26	0.30
Kavraisky VII	0.27	0.36	19.1	20.6	0.23	0.26
Putnins P3'	0.26	0.37	22.1	23.2	0.27	0.31
Winkel II	0.29	0.38	18.4	20.1	0.22	0.25
Eckert V	0.28	0.34	23.4	24.0	0.30	0.32
Wagner III	0.29	0.40	22.6	23.5	0.28	0.31
Winkel I	0.22	0.27	25.8	25.6	0.29	0.29
Eckert I	0.28	0.34	30.3	26.4	0.35	0.32
Eckert IV	0.00	0.00	28.7	31.4	0.35	0.43
Wagner IV = Putnins P2'	0.00	0.00	30.4	32.1	0.37	0.42
Werenskiold III	0.34	0.34	30.4	32.1	0.39	0.44
Wagner V	0.11	0.15	25.4	27.1	0.30	0.34
Putnins P4'	0.00	0.00	31.5	32.6	0.38	0.42
Werenskiold I	0.31	0.31	31.5	32.6	0.40	0.43
Flat-polar parabolic authalic	0.00	0.00	33.7	33.9	0.41	0.42
Flat-polar quartic authalic	0.00	0.00	32.1	33.2	0.39	0.42
Flat-polar sinusoidal authalic	0.00	0.00	35.1	34.6	0.42	0.42
Eckert VI	0.00	0.00	32.4	32.9	0.39	0.42
Wagner I = Kavraisky VI	0.00	0.00	31.9	32.7	0.38	0.42
Werenskiold II	0.30	0.30	31.9	32.7	0.40	0.43
Wagner II	0.11	0.15	26.9	27.7	0.31	0.34
Nell–Hammer	0.00	0.00	30.9	33.6	0.42	0.50
Eckert II	0.00	0.00	38.2	35.3	0.46	0.45
Robinson	0.21	0.25	21.4	23.2	0.26	0.30
Ginsburg VIII	0.49	0.59	20.3	17.5	0.29	0.29

Cylindrical projections	D_{ar}	D_{arc}	D_{an}	D_{anc}	D_{ab}	D_{abc}
Cylindrical equidistant:						
Plate *carrée*	0.55	0.77	16.8	21.0	0.27	0.38
Equi-rectangular projection	0.50	0.59	20.2	21.8	0.25	0.30
Cylindrical equal-area:						
Lambert	0.00	0.00	30.9	37.8	0.55	0.77
Behrmann	0.00	0.00	26.8	32.5	0.44	0.62
Peters	0.00	0.00	33.0	36.9	0.45	0.57
Pavlov	0.31	0.43	21.5	26.8	0.33	0.45
Mercator	3.40	5.45	0.0	0.0	0.55	0.77
Miller I	1.23	1.79	7.6	9.9	0.38	0.52
Miller II	0.93	1.33	10.8	13.8	0.34	0.47
Cylindrical stereographic:						
Braun	0.96	1.36	10.3	13.3	0.35	0.47
BSAM	0.73	1.03	9.0	11.6	0.28	0.38
Gall	0.70	0.89	10.6	12.5	0.29	0.36
Urmaev III	1.84	2.61	5.4	6.3	0.51	0.67

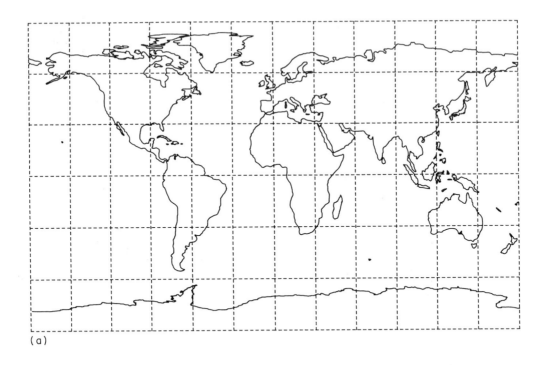

(a)

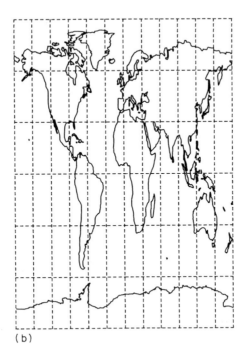

(b)

Fig. 4.1 Minimum error cylindrical equidistant projection: (a) minimization of D_{ab}; (b) minimization according to Airy's sleast squares criterion (equation 104, p. 334).

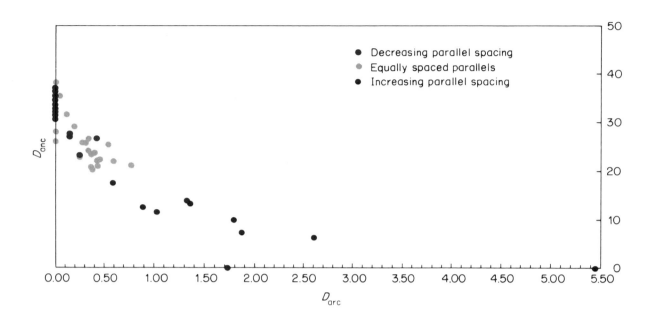

Fig. 4.2   Characteristic distortion parameters over continental area.

equal-area projections and projections with a small areal distortion generally have decreased parallel spacing. Graticules with equally spaced parallels occupy an intermediate position.

Equal-area and conformal projections always show extreme distortion values near the edges of the map. They are therefore not recommended for small-scale world mapping. Projections with small areal distortion are well suited for choropleth maps, those with small angular distortion are better suited for vector representations.

Projections which are neither conformal nor equal-area generally give a far better portrayal of the major continents. This can be inferred from their lower mean linear distortion values (Table 4.1). In particular the polyconic and pseudocylindrical projections with small areal distortion have favourable values for D_{ab} and D_{abc}. Kavraisky VII, Winkel II and the Winkel–Tripel projection have the lowest mean linear distortion. Nevertheless, if the conventional arrangement of the continents is not required oblique aspects of other projections can yield acceptable alternatives for these graticules (e.g. Briesemeister's projection, Atlantis projection). Finally one should realize that the quantitative analysis has its drawbacks and that other (qualitative) considerations are often important in the selection of an appropriate projection system for a particular application.

5

POLYCONIC PROJECTIONS

	Pole = point	Pole = line
Equally spaced parallels	5.1 Aitoff	5.2 Aitoff–Wagner 5.3 Winkel–Tripel
Decreasing parallel spacing	5.4 Hammer–Aitoff/Nordic/ Briesemeister 5.6 Eckert–Greifendorff	5.5 Hammer–Wagner
Increasing parallel spacing	5.7 Lambert 5.8 van der Grinten	

AITOFF PROJECTION

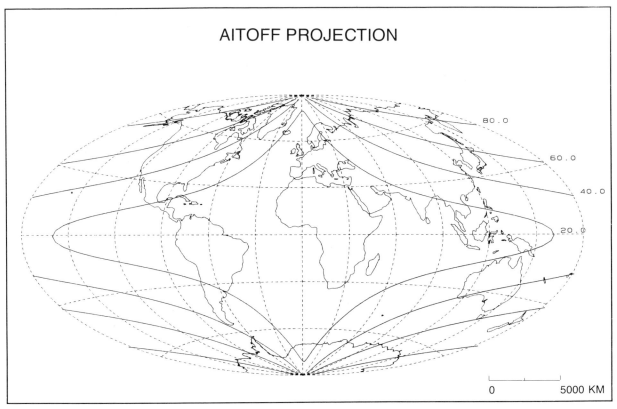

Lines of equal angular distortion

AITOFF PROJECTION

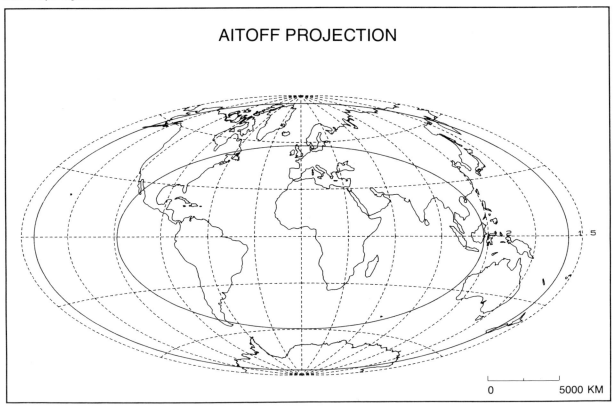

Lines of equal areal distortion

5.1. Polyconic projection with equally spaced parallels (Aitoff)

5.1.1 Normal Aspect

Other name(s): Aitoff projection

Author(s): David Aitoff (1889)

Description

By the end of the nineteenth century the Russian cartographer Aitoff developed a projection which he derived from the transverse aspect equidistant azimuthal projection. The construction principle involves the vertical projection of every point of the parent grid system on a plane which makes an angle of 60° and intersects the orginal projection along the central meridian.

This transformation has the effect of multiplying the *x*-coordinate of every point of the original graticule by a factor of two. At the same time the longitude λ is divided by the same factor in order to represent the total area of the world. Later this transformation inspired Hammer to develop his well-known equal-area projection (Hammer, 1892).

Aitoff's projection is neither equivalent nor conformal but offers a good compromise of both properties. It does not show the compression that occurs near the edges of Hammer's comparable equal-area projection. Moreover, due to the curvature of the parallels the intersection with the meridians is less sharp near the edges of the map than on a pseudocylindrical projection. This results in a more pleasing representation of the world.

A disadvantage of this projection is the compression in the higher latitudes that results from the meridional convergence. This is typical for each projection that shows the pole as a point. Later Wagner developed a graticule with pole line by generalizing Aitoff's transformation principle. This projection gives a more balanced representation of the earth (Aitoff–Wagner projection).

Transformation formulas

$$x = 2R\delta' \sin \lambda' \qquad\qquad y = -R\delta' \cos \lambda'$$

$$\delta' = \frac{\pi}{2} - \phi'$$

$$\sin \lambda' = \frac{\sin \frac{\lambda}{2} \cos \phi}{\cos \phi'} \qquad\qquad \cos \lambda' = -\frac{\sin \phi}{\cos \phi'}$$

$$\sin \phi' = \cos \phi \cos \frac{\lambda}{2}$$

where δ' = angular distance from the centre of projection ϕ', λ' = transformed latitude and longitude respectively (Section 2.2).

The transformation is indeterminate for the center of projection. However the coordinates of the centre need not be calculated explicitly ($x = 0$, $y = 0$). The coordinates of Aitoff's projection may also be calculated directly as a function of ϕ and λ. Substitution of the above expressions for the transformed latitude and longitude into the formulas for x and y leads to the following equations:

$$x = 2R \frac{\arccos\left(\cos \phi \, \cos \frac{1}{2}\lambda\right) \sin \frac{1}{2}\lambda}{\sqrt{\sin^2 \frac{1}{2}\lambda + \tan^2 \phi}} \qquad y = R \frac{\arccos\left(\cos \phi \, \cos \frac{1}{2}\lambda\right) \tan \phi}{\sqrt{\sin^2 \frac{1}{2}\lambda + \tan^2 \phi}}$$

Again these equations are indeterminate for the centre of projection.

Distortion characteristics

$D_{ar} = 0.23$	$D_{an} = 30.2$	$D_{ab} = 0.36$
$D_{arc} = 0.19$	$D_{anc} = 28.9$	$D_{abc} = 0.34$

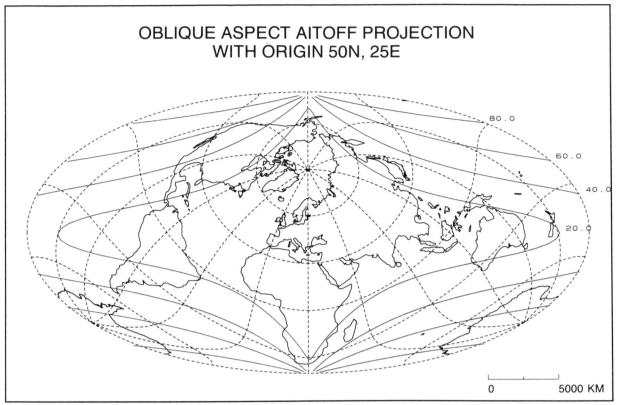

Lines of equal angular distortion

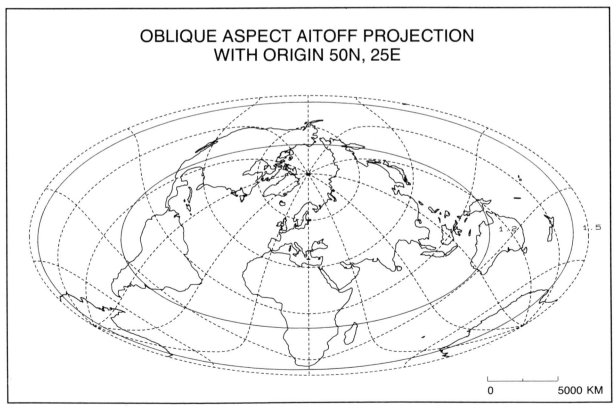

Lines of equal areal distortion

5.1 Polyconic projection with equally spaced parallels (Aitoff) (CONTINUED)

5.1.2 Oblique Aspect

Description

Since Aitoff's projection represents the total surface of the earth within an ellipse it is well suited for the development of oblique aspects. Taking a central meridian as smaller axis, the minimization of the mean scale distortion over the continental area D_{abc} leads to an oblique aspect Aitoff projection with center 60°N, 25°E. This aspect cuts the southern part of Africa and may therefore not be considered as a satisfying solution. However, a small displacement of the center of projection from the optimal solution to 50°N, 25°E solves the problem.

A comparison of this projection with the oblique aspect Hammer projection with the same center shows that the latter is less suited because of the considerable compression near the edges of the map (compare the representation of Africa on both projections).

If the equal-area property is not required the oblique aspect Aitoff projection may be considered as a good alternative for the more conventional equatorial projections in most contemporary atlases.

Distortion characteristics

The minimum-error oblique Aitoff projection ($\phi_c = 60°$N, $\lambda_c = 25°$E) gives the following values for the distortion over continental area:

$$D_{arc} = 0.14 \qquad D_{anc} = 15.8 \qquad D_{abc} = 0.17$$

A small displacement of the centre of projection ($\phi_c = 50°$N, $\lambda_c = 25°$E) has little effect on these values.

$$D_{arc} = 0.14 \qquad D_{anc} = 16.6 \qquad D_{abc} = 0.18$$

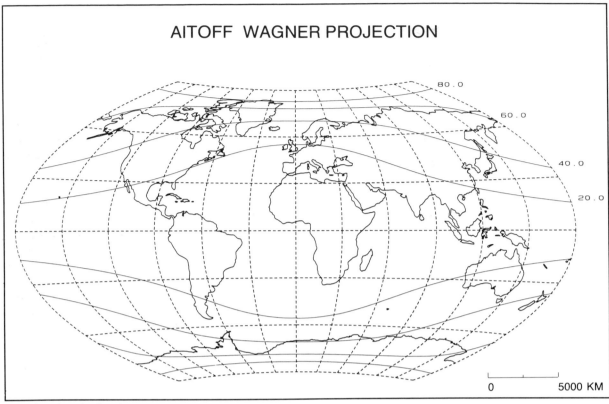

Lines of equal angular distortion

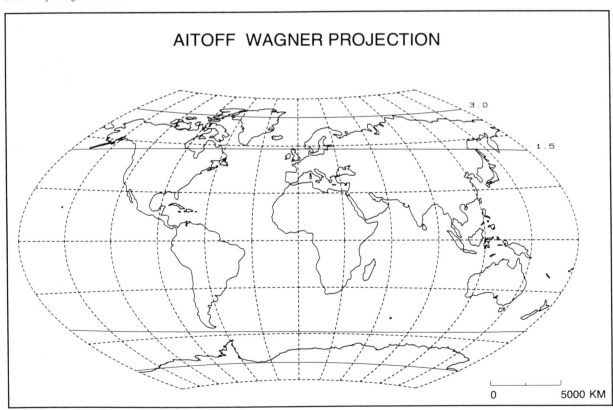

Lines of equal areal distortion

5.2 Polyconic projection with equally spaced parallels and pole line (Wagner)

Other name(s):   Aitoff–Wagner projection

Author(s):   Karlheinz Wagner (1962)

Description

The application of Wagner's transformation principle (Section 2.4) to the transverse aspect azimuthal equidistant projection leads to a projection system that allows for the development of an unlimited number of projections depending on the values of three parameters m, n, k in the general transformation formulas. For $m = 1$, $n = 0.5$ and $k = \sqrt{2}$ Aitoff's graticule is obtained. The E–W compression of Aitoff's projection may be avoided by the introduction of a pole line. Wagner proposed a redefinition of m, n and k to obtain a projection whereby the length of the pole line and the curvature of the parallels is comparable with the Winkel–Tripel projection, which is a polyconic projection with very favourable distortion patterns. The resulting graticule shows a great resemblance to Winkel's projection only the latter has a straight pole line. The angular and scale distortion of the Aitoff–Wagner projection is low. On the other hand the areal distortion is relatively high.

Transformation formulas

$$x = R \frac{k}{\sqrt{nm}} \delta' \sin \lambda' \qquad\qquad y = -R \frac{1}{k\sqrt{nm}} \delta' \cos \lambda'$$

$$\delta' = \frac{\pi}{2} - \phi' \qquad \sin \lambda' = \frac{\sin(n\lambda) \cos(m\phi)}{\cos \phi'} \qquad \cos \lambda' = -\frac{\sin(m\phi)}{\cos \phi'}$$

$$\sin \phi' = \cos(m\phi) \cos(n\lambda)$$

$$m = \frac{7}{9} \qquad\qquad n = \frac{5}{18} \qquad\qquad k = \sqrt{\frac{14}{5}}$$

where δ' = angular distance from the centre of projection, and ϕ', λ', = transformed latitude and longitude respectively (Section 2.2).

The transformation is indeterminate for the center of projection. However, the coordinates of the center need not be calculated explicitly ($x = 0$, $y = 0$).

Distortion characteristics

$D_{ar} = 0.33$	$D_{an} = 21.2$	$D_{ab} = 0.26$
$D_{arc} = 0.42$	$D_{anc} = 22.0$	$D_{abc} = 0.29$

WINKEL—TRIPEL PROJECTION

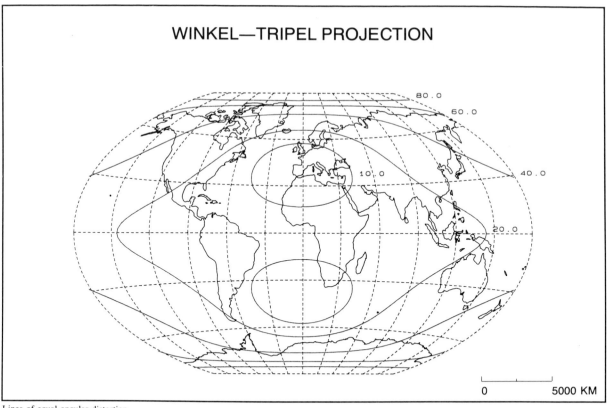

Lines of equal angular distortion

WINKEL—TRIPEL PROJECTION

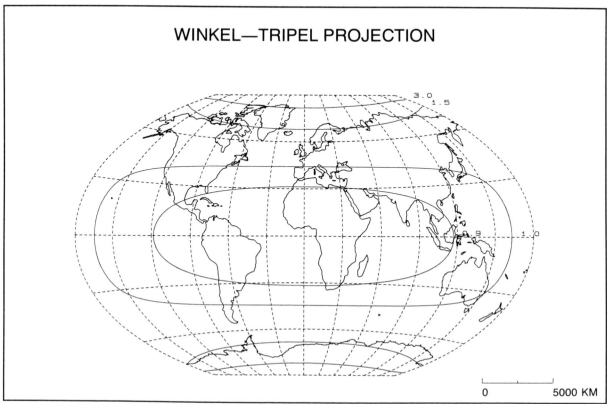

Lines of equal areal distortion

5.3   Polyconic projection with equally spaced parallels and pole line (Winkel)

Other name(s):   Winkel–Tripel projection

Author(s):   Oswald Winkel (1928, 1939)

Description

O. Winkel developed three projections which are the arithmetic mean of the cylindrical equidistant projection and a projection which represents the pole as a point. The best known is his third projection which is the arithmetic mean of Aitoff's projection and the cylindrical equidistant projection with two standard parallels. To calculate the standard latitude Winkel gives the following expression:

$$k = \cos \phi_0 = \frac{\sin \phi_1}{\phi_1}$$

with ϕ_1 indicating the outer parallel of the zone which is to be represented. Winkel developed his projection for the representation of the *oikumene* (inhabited world) which he limited to $\phi = \pm 70°$. In this case ϕ_0 becomes $39.7°$. For the representation of the whole world ($\phi_1 = 90°$) the above formula leads to $\phi_0 = 50.5°$. This value approximates the standard latitude that is obtained by minimizing D_{ar}.

The Winkel–Tripel projection has slightly curved parallels and a straight pole line shorter than or half the length of the equator (depending on whether $k < 1$ or $k = 1$). It gives a satisfying representation of the world and has well-balanced distortion patterns. The mean scale distortions D_{ab} and D_{abc} have a very low value. Compared with the Aitoff–Wagner projection the areal distortion on Winkel's graticule is much lower.

The Winkel–Tripel projection may be considered as one of the best suited projections for general-purpose world maps and receives a lot of attention in contemporary atlas cartography.

Transformation formulas

$$x = \frac{R}{2}\left[k\lambda + \frac{2\arccos\left(\cos\phi \, \cos\frac{1}{2}\lambda\right)\sin\frac{1}{2}\lambda}{\sqrt{\sin^2\frac{1}{2}\lambda + \tan^2\phi}} \right]$$

$$y = \frac{R}{2}\left[\phi + \frac{\arccos\left(\cos\phi \, \cos\frac{1}{2}\lambda\right)\tan\phi}{\sqrt{\sin^2\frac{1}{2}\lambda + \tan^2\phi}} \right]$$

$$k = \cos \phi_0$$

Distortion characteristics

$D_{ar} = 0.17$	$D_{an} = 23.4$	$D_{ab} = 0.25$
$D_{arc} = 0.25$	$D_{anc} = 22.7$	$D_{abc} = 0.26$

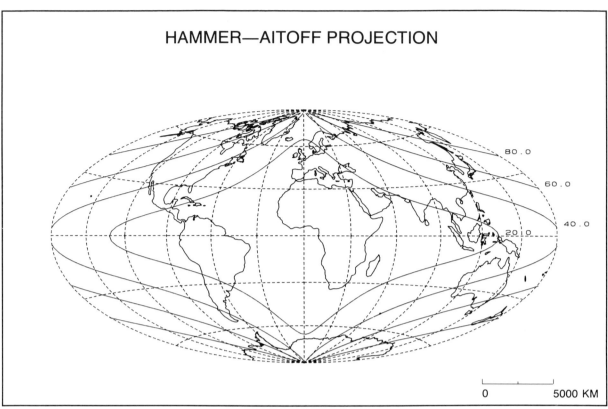

HAMMER—AITOFF PROJECTION

80.0
60.0
40.0
20 0

0   5000 KM

Lines of equal angular distortion

5.4 Polyconic equal-area projection (Hammer)

5.4.1 Normal Aspect

Other name(s): Hammer–Aitoff projection

Author(s): E. Hammer (1892)

Description

Hammer realized that Aitoff's idea to develop a world map projection by transformation of the transverse azimuthal equidistant projection (Aitoff projection) could be applied to each projection that represents a hemisphere as a circle. He proved that Aitoff's transformation is equivalent and applied it to the azimuthal equal-area projection. The resulting equivalent graticule is named after him. It is very similar to Aitoff's projection, but as a result of the equal-area property compression near the edges of the map is inevitable.

Hammer states that the angular distortion is less on his projection than on Mollweide's graticule. The study of Behrmann (Section 3.1.1) as well as our own analysis show that this statement is to be interpreted with some reserve. Only integration of distortion over continental area leads to slightly better distortion values for the Hammer–Aitoff projection due to the fact that parallels and meridians intersect at a less sharp angle in higher latitudes than on Mollweide's graticule.

Transformation formulas

$$x = 4R \, \sin \frac{\delta'}{2} \, \sin \lambda' \qquad\qquad y = -2R \, \sin \frac{\delta'}{2} \, \cos \lambda'$$

$$\delta' = \frac{\pi}{2} - \phi' \qquad \sin \lambda' = \frac{\sin \frac{1}{2}\lambda \, \cos \phi}{\cos \phi'} \qquad \cos \lambda' = -\frac{\sin \phi}{\cos \phi'}$$

$$\sin \phi' = \cos \phi \, \cos \frac{1}{2}\lambda$$

where δ' = angular distance from the centre of projection, and ϕ', λ' = transformed latitude and longitude respectively (Section 2.2).

The transformation is indeterminate for the center of projection. However, the coordinates of the centre need not be calculated explicitly ($x = 0$, $y = 0$). The coordinates of Hammer's projection may also be calculated directly as a function of ϕ and λ. Substitution of the above expressions for the transformed latitude and longitude into the formulas for x and y leads to the following equations:

$$x = 2R\sqrt{2} \, \frac{\cos \phi \, \sin \frac{1}{2}\lambda}{\sqrt{1 + \cos \phi \, \cos \frac{1}{2}\lambda}} \qquad\qquad y = R\sqrt{2} \, \frac{\sin \phi}{\sqrt{1 + \cos \phi \, \cos \frac{1}{2}\lambda}}$$

Distortion characteristics

$D_{ar} = 0.00$	$D_{an} = 35.7$	$D_{ab} = 0.43$
$D_{arc} = 0.00$	$D_{anc} = 33.6$	$D_{abc} = 0.41$

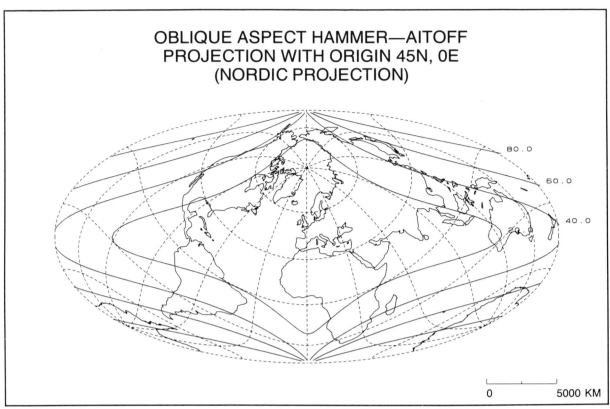

**OBLIQUE ASPECT HAMMER—AITOFF
PROJECTION WITH ORIGIN 45N, 0E
(NORDIC PROJECTION)**

80.0

60.0

40.0

0   5000 KM

Lines of equal angular distortion

5.4   Polyconic equal-area projection (Hammer) (CONTINUED)

5.4.2   *Oblique Aspect*

As the Hammer–Aitoff projection represents the world within an ellipse it lends itself well to the development of oblique aspects. As described in Section 2.2, a change of aspect has no influence on the distortion pattern of the projection. It is only recentered with respect to the earth's surface. Therefore the global distortion parameters are not altered. The transformation formulas of an oblique aspect projection may be derived from those of the normal aspect by applying the rules of spherical trigonometry. This procedure is described in Section 2.2.

Two examples of oblique aspect Hammer–Aitoff projections are familiar in atlas cartography : Bartholomew's Nordic projection and Briesemeister's projection.

1.   The Nordic projection

Author(s):   John Bartholomew

Description

This projection was developed by Bartholomew and is an oblique aspect of the Hammer–Aitoff projection with as smaller axis the meridians of 0° and 180° and as greater axis a great circle that reaches its vertex in the centre of the projection at 0°, 45°N. Apart from being rotated over an angle of 45° the distortion pattern of this projection is identical to that of the normal aspect Aitoff projection. Bartholomew developed the projection in order to obtain an optimal representation of Europe and of routes in the Atlantic, Arctic and Indian Oceans. The projection is less suited if a good representation of the overall continental area is required. The distortion parameters are substantially lower than for the normal aspect of the projection because the center is situated in the middle of the continental areas of the Northern Hemisphere. Nevertheless the continents of the Southern Hemisphere are severely distorted.

Distortion characteristics	$D_{ar} = 0.00$	$D_{an} = 35.7$	$D_{ab} = 0.43$
	$D_{arc} = 0.00$	$D_{anc} = 23.7$	$D_{abc} = 0.26$

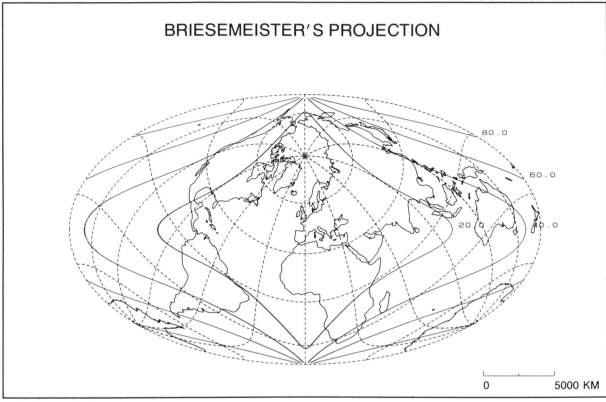

BRIESEMEISTER'S PROJECTION

0   5000 KM

Lines of equal angular distortion

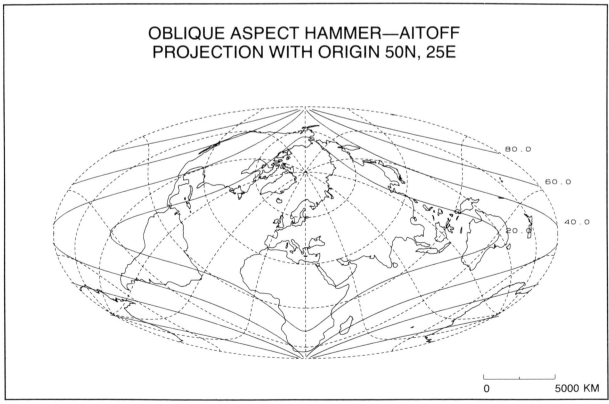

OBLIQUE ASPECT HAMMER—AITOFF
PROJECTION WITH ORIGIN 50N, 25E

0   5000 KM

Lines of equal angular distortion

5.4 Polyconic equal-area projection (Hammer) (CONTINUED)

5.4.2 Oblique Aspect

2. Briesemeister's projection

Author(s): William Briesemeister (1953)

Description

W. Briesemeister developed a modification of the oblique aspect Hammer–Aitoff projection with center at 45° N. He noticed that the parallels near the North Pole are ovals on this projection and argued that a better projection would be obtained by making these parallels circular. Therefore he changed the ratio of the axes so that the greater axis becomes 1.75 times the smaller axis and the equal-area property is maintained.

The transformation formulas of Briesemeister's projection are almost identical to those of the oblique aspect Hammer–Aitoff projection, only the x-coordinate is multiplied by a factor of $\sqrt{1.75/2}$, while the y-coordinate is divided by the same factor. Originally Briesemeister situated the center of the projection at 10° E. We placed it at 0° longitude to allow for a better comparison with the Nordic projection.

Briesemeister's projection partly compensates the N–S compression of the continents in the Southern Hemisphere. Therefore it is better suited for global representations of the world than the Nordic projection. It forms an alternative for the oblique aspect Aitoff projection if the equal-area property is required.

Transformation formulas

$$x = 2R\sqrt{2}m\,\frac{\cos\phi'\,\sin\frac{1}{2}\lambda'}{\sqrt{1 + \cos\phi'\,\cos\frac{1}{2}\lambda'}} \qquad\qquad y = \frac{R\sqrt{2}}{m}\,\frac{\sin\phi'}{\sqrt{1 + \cos\phi'\,\cos\frac{1}{2}\lambda'}}$$

$$m = \sqrt{\frac{A}{2B}} = \sqrt{\frac{1.75}{2}}$$

ϕ', λ' = transformed latitude and longitude respectively (Section 2.2), A = length of the greater axis and B = length of the smaller axis.

The transformed latitude and longitude of a point on the globe are obtained by equations (63), (64) and (65) in Section 2.2.

Distortion characteristics

$D_{ar} = 0.00$	$D_{an} = 35.8$	$D_{ab} = 0.47$
$D_{arc} = 0.00$	$D_{anc} = 24.0$	$D_{abc} = 0.26$

3. Other oblique aspects

Description

Apart from these two well-known projections it is possible to create an unlimited number of other oblique aspect Hammer–Aitoff projections, with or without modification of the ratio of the axes. Taking a central meridian as smaller axis, the minimization of the distortion parameter D_{abc} leads to a centre at 25° E, 60° N. Hence the southern part of Africa is split up. This may be avoided by moving the centre of projection slightly to the south. Nevertheless an oblique Hammer–Aitoff projection with centre at 25° E, 50° N does not give a satisfying representation of the continental areas although Africa is represented without interruption. This is caused by the distortion pattern of the Hammer–Aitoff projection which is characterized by a considerable compression near the edges of the map. This problem can only be avoided by giving up the equal-area property. The optimized oblique aspect Aitoff projection gives a much better representation of the continental areas and illustrates the potential of experimentation with oblique aspects of a projection.

Distortion characteristics

The minimum-error oblique Hammer–Aitoff projection ($\phi_c = 60°$ N, $\lambda_c = 25°$ E) has the following distortion parameters.

$$D_{arc} = 0.00 \qquad D_{anc} = 19.3 \qquad D_{abc} = 0.21$$

A movement of the projection centre to 25° E, 50° N leads to a small change in the values of these parameters.

$$D_{arc} = 0.0 \qquad D_{anc} = 20.1 \qquad D_{abc} = 0.22$$

ECKERT—GREIFENDORFF'S PROJECTION

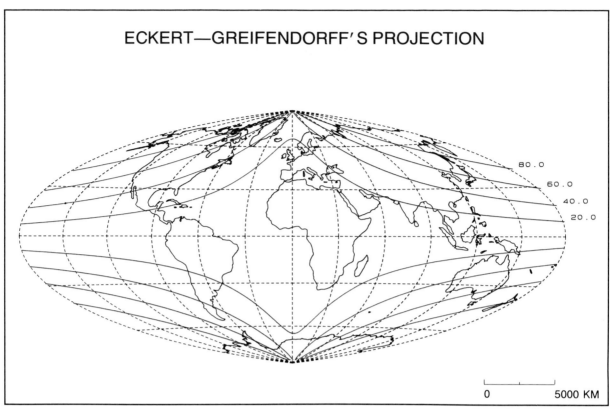

80.0
60.0
40.0
20.0

0   5000 KM

Lines of equal angular distortion

5.5 Polyconic equal-area projection (Eckert-Greifendorff)

Other name(s): Eckert-Greifendorff projection

Author(s): Max Eckert-Greifendorff (1935)

Description

Almost 30 years after the publication of his six pseudocylindrical projections (1906) M. Eckert developed a polyconic equal-area projection. This was inspired by D. Aitoff's idea to transform a projection which represents a hemisphere by a circle in such a way that it becomes possible to represent the whole world. This transformation involves the multiplication of the *x*-coordinate of the original graticule by a chosen factor and the division of the geographical longitude in the transformation formulas by the same factor (Aitoff projection). E. Hammer proved that Aitoff's transformation has no effect on the areal distortion. He applied it to the transverse aspect azimuthal equivalent projection and obtained an equal-area graticule that represents the whole world within an ellipse (Hammer–Aitoff projection). Just like Aitoff he carried out the transformation with a factor of two.

M. Eckert also started from the transverse aspect azimuthal equivalent projection but, since he wanted the parallels to approximate straight lines, he applied Aitoff's transformation with a factor of four. This means that in the graticule which is obtained after multiplication of the *x*-coordinate by this factor the meridian of 45° becomes the meridian of 180°, the meridian of 30° becomes the meridian of 120° and so on. The resulting equal-area projection resembles a pseudocylindrical projection due to the very slight curvature of the parallels. Since in the parent projection the spacing of the meridians along the equator decreases towards the edges of the map, the ratio of the axes is not correct ($4/\sqrt{2}$ instead of $2/1$). Moreover the outer meridian is not an ellipse as on Hammer's graticule. A discontinuity appears at the poles of the projection.

Transformation formulas

$$x = 8R \, \sin \frac{\delta'}{2} \, \sin \lambda' \qquad\qquad y = -2R \, \sin \frac{\delta'}{2} \, \cos \lambda'$$

$$\delta' = \frac{\pi}{2} - \phi' \qquad \sin \lambda' = \frac{\sin \frac{1}{4}\lambda \, \cos \phi}{\cos \phi'} \qquad \cos \lambda' = -\frac{\sin \phi}{\cos \phi'}$$

$$\sin \phi' = \cos \phi \, \cos \frac{\lambda}{4}$$

where δ' = angular distance from the centre of projection, and ϕ', λ' = transformed latitude and longitude respectively (Section 2.2).

The transformation is indeterminate for the centre of projection. However, the coordinates of the centre need not be calculated explicitly ($x = 0$, $y = 0$). The coordinates may also be calculated directly as a function of ϕ and λ. Substitution of the above expressions for the transformed latitude and longitude into the formulas for x and y leads to the following equations:

$$x = 4R\sqrt{2} \, \frac{\cos \phi \, \sin \frac{1}{4}\lambda}{\sqrt{1 + \cos \phi \, \cos \frac{1}{4}\lambda}} \qquad y = R\sqrt{2} \, \frac{\sin \phi}{\sqrt{1 + \cos \phi \, \cos \frac{1}{4}\lambda}}$$

Distortion characteristics

$D_{ar} = 0.00$	$D_{an} = 35.5$	$D_{ab} = 0.45$
$D_{arc} = 0.00$	$D_{anc} = 35.6$	$D_{abc} = 0.46$

HAMMER—WAGNER PROJECTION

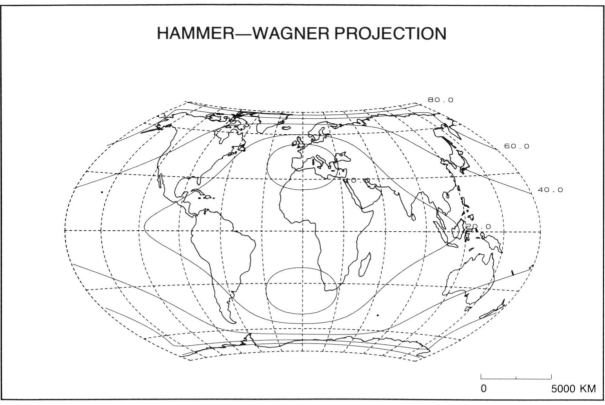

0   5000 KM

Lines of equal angular distortion

HAMMER—WAGNER PROJECTION WITH
CENTRAL MERIDIAN 150E

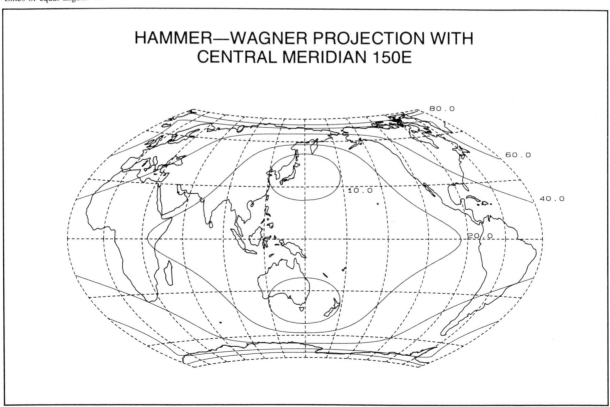

Lines of equal angular distortion

5.6 Polyconic equal-area projection with pole line (Wagner)

Other name(s): Hammer–Wagner projection

Author(s): Karlheinz Wagner (1962)

Description

As mentioned earlier, Aitoff's projection (polyconic projection with equally spaced parallels) is a special case of the general projection system which originates from the application of Wagner's transformation principle (Section 2.4) to the transverse aspect azimuthal equidistant projection.

In the same way Hammer's projection (polyconic equal-area projection) is a special case of Wagner's general projection system which is obtained by the application of his transformation to the transverse aspect azimuthal equivalent projection. Hammer's projection is characterized by a considerable compression near the edges of the map as a consequence of the increasing curvature of the parallels. By redefinition of the parameters m, n and k in the transformation formulas of the general projection system (for Hammer's projection $m = 1$, $n = 0.5$, $k = \sqrt{2}$) a graticule with a more

favourable distortion pattern can be obtained.

Wagner proposed a projection with a pole line approximately half the length of the equator, slightly curved parallels and a 2/1 ratio of the axes. This so-called Hammer–Wagner projection gives a more familiar representation of the continents than Hammer's graticule and the pseudocylindrical equal-area projections. The distortion parameters are lower than for all equal-area projections of the polyconic and pseudocylindrical class which are treated in this directory. As Wagner stated, the projection may be considered as the equal-area equivalent of the Winkel–Tripel projection (polyconic projection with equally spaced parallels and pole line). Therefore it deserves a broader attention in cartographic applications that require the equal-area property.

Transformation formulas

$$x = 2R \frac{k}{\sqrt{nm}} \sin \frac{\delta'}{2} \sin \lambda' \qquad\qquad y = -2R \frac{1}{k\sqrt{nm}} \sin \frac{\delta'}{2} \cos \lambda'$$

$$\delta' = \frac{\pi}{2} - \phi' \qquad\qquad \sin \lambda' = \frac{\sin(n\lambda)\cos\psi}{\cos\phi'} \qquad \cos\lambda' = -\frac{\sin\psi}{\cos\phi'}$$

$$\sin\phi' = \cos(n\lambda)\cos\psi$$

$$\sin\psi = m\sin\phi$$

$$m = 0.9063 \qquad\qquad n = 0.3333 \qquad\qquad k = 1.4660$$

where δ' = angular distance from the centre of projection, and ϕ', λ' = transformed latitude and longitude respectively (Section 2.2).

The transformation is indeterminate for the centre of projection. However, the coordinates of the centre need not be calculated explicitly ($x = 0$, $y = 0$).

Distortion characteristics

$D_{ar} = 0.00$	$D_{an} = 30.7$	$D_{ab} = 0.36$
$D_{arc} = 0.00$	$D_{anc} = 30.7$	$D_{abc} = 0.38$

Recentered Hammer–Wagner projection with central meridian at 150° E

Alternative views can be obtained by focusing on other parts of the world. This can easily be accomplished through a repositioning of the central meridian. Choosing the latter at

150° E yields a map which centers on eastern Asia and Australia without intersecting the major continents.

LAMBERT'S CONFORMAL PROJECTION
OF THE WORLD IN A CIRCLE

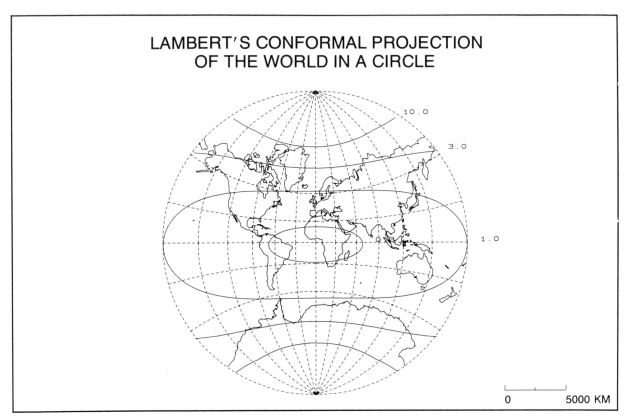

Lines of equal areal distortion

5.7 Polyconic conformal projection with circular meridians and parallels (Lambert)

Other name(s):	Lambert's conformal projection of the world in a circle/Lagrange's projection
Author(s):	J. H. Lambert (1772)

Description

J. H. Lambert (1728–77) made many contributions to the study of map projections. In a famous book on the subject of mathematical cartography (1772) he described a conformal projection of the world within a circle. The projection is sometimes wrongly attributed to Lagrange. Meridians and parallels are represented by arcs of circles and the spacing along the equator and central meridian increases from the center to the edge of the map. This is a necessary condition to assure the property of conformality. Although Lambert only described the projection mathematically and did not pay any attention to practical considerations, A. van der Grinten (1904) described the relatively simple geometrical construction of the graticule.

The distortion pattern of the projection shows that areas near the edges of the map are considerably enlarged with respect to areas near the center of the projection. The scale variation is extreme in the N–S direction. This results from the $1/1$ ratio of the axes and the conformality.

van der Grinten gave up the absolute conformality to obtain a more uniform distribution of distortion. On his first projection (van der Grinten I) the equator is equally divided by the meridians while the parallels are less curved than on Lambert's projection. In this way the representation of the continents is more pleasing.

Transformation formulas

$$x = \frac{2R\sqrt{1 - \tan^2 \tfrac{1}{2}\phi}\ \sin \tfrac{1}{2}\lambda}{1 + \sqrt{1 - \tan^2 \tfrac{1}{2}\phi}\ \cos \tfrac{1}{2}\lambda}$$

$$y = \frac{2R\ \tan \tfrac{1}{2}\phi}{1 + \sqrt{1 - \tan^2 \tfrac{1}{2}\phi}\ \cos \tfrac{1}{2}\lambda}$$

Distortion characteristics

$D_{ar} = 1.38$	$D_{an} = 0.0$	$D_{ab} = 0.49$
$D_{arc} = 1.73$	$D_{anc} = 0.0$	$D_{abc} = 0.59$

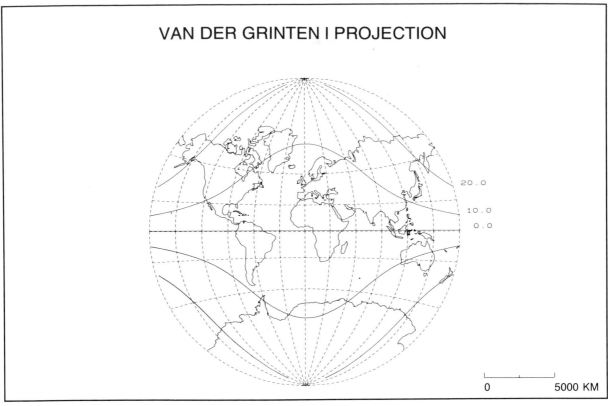

VAN DER GRINTEN I PROJECTION

20.0
10.0
0.0

0   5000 KM

Lines of equal angular distortion

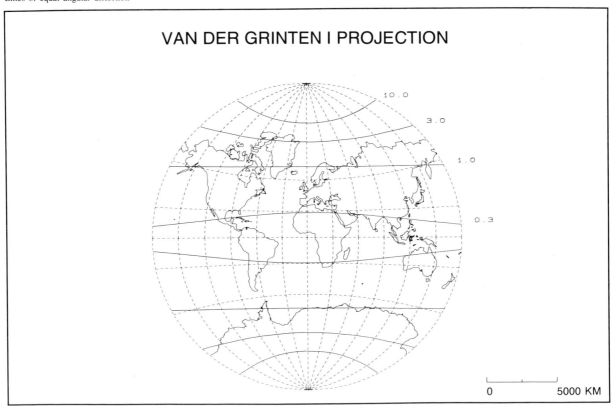

VAN DER GRINTEN I PROJECTION

10.0
3.0
1.0
0.3

0   5000 KM

Lines of equal areal distortion

5.8 Polyconic projection with circular meridians and parallels (van der Grinten)

Other name(s):   van der Grinten I projection

Author(s):   Alphons J. van der Grinten (1904)

Description

In 1904 A. J. van der Grinten proposed four new projections for world maps. The first three projections represent the world within a circle, while the fourth encloses the world within two intersecting circles. The projections are neither conformal nor equal-area. Van der Grinten's purpose was to develop projections which were graphically easy to construct. Therefore meridians and parallels had to be arcs of circles or straight lines. According to van der Grinten the use of arcs of circles for the meridians and the parallels generally leads to an apple-shaped outer meridian (two intersecting circles). The central meridian and the equator are straight lines. When both intersecting circles coincide the outline of the projection becomes a circle. Van der Grinten states that in this case distortion is minimal. He criticizes Lambert's conformal projection of the world within a circle. This projection shows a substantial areal enlargement from the centre of the projection to the outer meridian. The scale variation along the equator

and central meridian is considerable. Van der Grinten's three projections of the world within a circle have a constant scale along the equator. They differ from each other in the curvature of the parallels. His most famous projection, the so-called van der Grinten I, is characterized by a gradual increase of distortion from the equator to the poles. As a consequence the continents, with exception from the polar areas, are represented well. Therefore most world maps on this projection only reach the northern coast of Greenland and the Antarctic coast.

In the original paper van der Grinten emphasizes the graphical construction principle of the projection. He does not give transformation formulas. More recently mathematical formulas were published by O'Keefe and Greenberg (1977) and by Snyder (1979, 1982). The transformation formulas given below are those developed by Snyder. In spite of the relatively simple graphical construction the formulas are rather complex and only manageable by computer.

Transformation formulas

$$x = \pm r \frac{A(G - P^2) + \sqrt{A^2(G - P^2)^2 - (P^2 + A^2)(G^2 - P^2)}}{P^2 + A^2}$$

$$y = \pm r \sqrt{1 - \left(\frac{x}{r}\right)^2 - 2A \left|\frac{x}{r}\right|}$$

where

$$A = \frac{1}{2} \left| \frac{\pi}{\lambda} - \frac{\lambda}{\pi} \right|$$

$$P = G \left(\frac{2}{\sin \theta} - 1 \right)$$

$$G = \frac{\cos \theta}{\sin \theta + \cos \theta - 1}$$

$$\theta = \arcsin \left| \frac{2\phi}{\pi} \right|$$

The x-coordinate takes the sign of λ, the y-coordinate the sign of ϕ. For $r = R\pi$ the equator is represented in correct length, for $r = R\pi/2$ the central meridian is of true length. For

the calculation of the distortions and the creation of the maps the present authors chose $r = R\pi/2$. If $\phi = 0$ or $\lambda = 0$ the above equations are indeterminant.

For $\phi = 0$ $x = r \frac{\lambda}{\pi}$ $y = 0$

and for $\lambda = 0$, $x = 0$ $y = \pm r \tan \left(\frac{\theta}{2} \right)$

the y-coordinate taking the sign of ϕ.

Distortion characteristics $D_{ar} = 1.87$ $D_{an} = 7.7$ $D_{ab} = 0.67$

$D_{arc} = 1.87$ $D_{anc} = 7.3$ $D_{abc} = 0.67$

6

PSEUDOCYLINDRICAL PROJECTIONS

	Pole = point	Pole = line
Equally spaced parallels	6.1 Apianus II/Arago	6.5 Ortelius
	6.2 Putnins P1	6.6 Eckert III
	6.3 Putnins P3	6.7 Wagner VI/Putnins P1′
	6.4 Sanson	6.8 Kavraisky VII
		6.9 Putnins P3′
		6.10 Winkel II
		6.11 Eckert V
		6.12 Wagner III
		6.13 Winkel I
		6.14 Eckert I
Decreasing parallel spacing	6.15 Mollweide/Atlantis	6.21 Eckert IV
	6.16 Putnins P2	6.22 Wagner IV = Putnins P2′/Werenskiold III
	6.17 Boggs	6.23 Wagner V
	6.18 Craster = Putnins P4	6.24 Putnins P4′/Werenskiold I
	6.19 Adams	6.25 Flat-polar parabolic authalic
	6.20 Kavraisky V	6.26 Flat-polar quartic authalic
		6.27 Flat-polar sinusoidal authalic
		6.28 Eckert VI
		6.29 Wagner I
		6.30 Wagner II
		6.31 Nell-Hammer
		6.32 Eckert II
		6.33 Robinson
Increasing parallel spacing		6.34 Ginsburg VIII

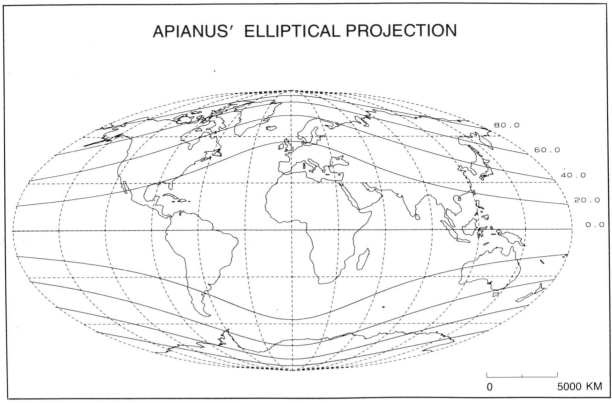

Lines of equal angular distortion

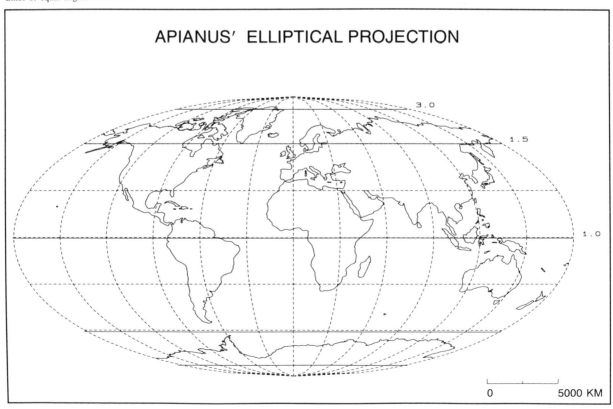

Lines of equal areal distortion

6.1 Pseudocylindrical projection with equally spaced parallels and elliptical meridians (Apianus)

Other name(s): Apianus' elliptical projection/Apianus II projection/
Arago's projection

Author(s): Petrus Apianus (Peter Bienewitz)

Description

The inner hemisphere of this projection is represented as a circle with radius r. The parallels are equally spaced along the central meridian and equally divided by the elliptical meridians. When this construction principle is extended outside the inner hemisphere it becomes possible to represent the whole world.

Depending on the value of the radius r two variants of this projection may be distinguished from one another. If the principal scale is preserved along the central meridian and the equator Apianus's projection is obtained. If, on the other hand, the area of the planisphere equals the area of the generating globe the projection is attributed to Arago. On this projection the area enclosed by two meridians is also preserved. The modification of the radius r only involves an equal scaling in the $x-$ and the y-direction and does not influence the angular distortion of the projection.

Transformation formulas

$$x = \frac{2\lambda}{\pi} \sqrt{r^2 - y^2}$$

$$y = 2r\frac{\phi}{\pi}$$

$$r = R\frac{\pi}{2} \qquad \text{Apianus's second projection}$$

$$r = R\sqrt{2} \qquad \text{Arago's projection}$$

$$r = \text{radius of the inner hemisphere}$$

Distortion characteristics

Apianus II	$D_{ar} = 0.23$		$D_{an} = 24.2$		$D_{ab} = 0.30$
	$D_{arc} = 0.31$		$D_{anc} = 25.5$		$D_{abc} = 0.33$
Arago	$D_{ar} = 0.21$		$D_{an} = 24.2$		$D_{ab} = 0.31$
	$D_{arc} = 0.25$		$D_{anc} = 25.5$		$D_{abc} = 0.34$

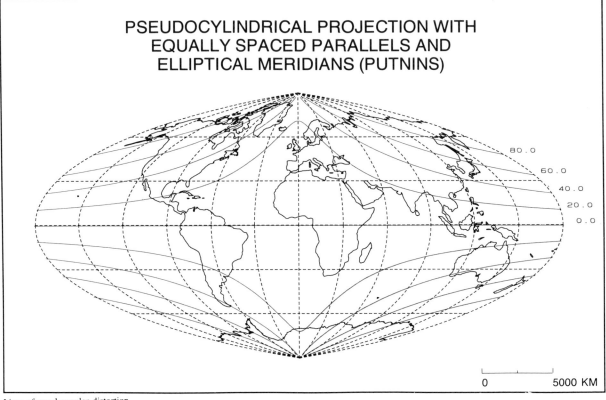

PSEUDOCYLINDRICAL PROJECTION WITH EQUALLY SPACED PARALLELS AND ELLIPTICAL MERIDIANS (PUTNINS)

Lines of equal angular distortion

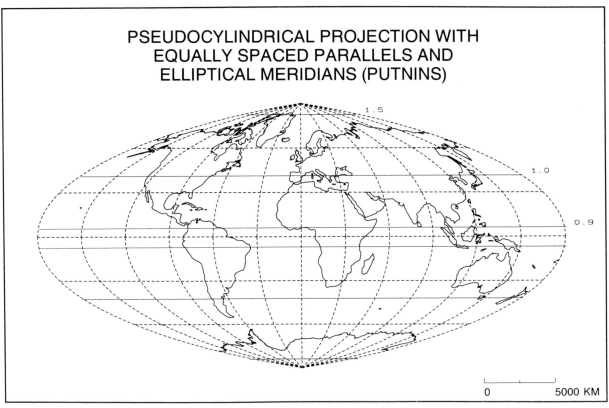

PSEUDOCYLINDRICAL PROJECTION WITH EQUALLY SPACED PARALLELS AND ELLIPTICAL MERIDIANS (PUTNINS)

Lines of equal areal distortion

6.2 Pseudocylindrical projection with equally spaced parallels and elliptical meridians (Putnins)

Other name(s):   Putnins P1 projection

Author(s):   Reinholds V. Putnins (1934)

Description

In 1934 R. V. Putnins proposed twelve conic section pseudocylindrical projections, four with elliptical meridians (P1, P1', P2, P2'), four with parabolic meridians (P3, P3', P4, P4') and four with hyperbolic meridians (P5, P5', P6, P6'). The uneven numbering corresponds to projections with equally spaced parallels, while the even numbered projections are equal-area. The quote in the codification designates the presence of a pole line. An overview of all combinations leading to Putnins' twelve projections is given in Table 6.1. The structure of Table 6.1 is in conformity with the general classification scheme which underlies this directory of projection systems.

Table 6.1   Overview of Putnins' twelve projections

	Equally spaced parallels		Decreasing parallel spacing (equal-area)	
	Pole = point	Pole = line	Pole = point	Pole = line
Elliptical meridians	P1	P1'	P2	P2'
Parabolic meridians	P3	P3'	P4	P4'
Hyperbolic meridians	P5	P5'	P6	P6'

The projections with hyperbolic meridians will not be discussed here since these meridians meet in the pole at an angle which is even sharper than is the case for projections with sinusoidal meridians. Putnins' elliptical and parabolic projections on the other hand are all incorporated in this directory, although not all under this heading. Three of them (P1', P2' and P4) are identical to graticules developed by other authors which are better known under their names (for reference see index).

Putnins' elliptical projections only use a fixed portion of a semiellipse for a given meridian which explains the conspicuous polar discontinuity. This is the main difference between Putnins P1 and Apianus II, which uses complete semiellipses for meridians. Apart from that distinction Putnins P1 is constructed according to the same principle as Apianus II. The ratio of the axes is correct, the parallels are equally spaced and are equally divided by the elliptical meridians. The scale factor along the axes of the graticule is reduced to make the total area equal to the area of the generating globe (similar to Arago's version of Apianus II). Because of the nature of the meridians the graticule of Putnins P1 is about halfway between Apianus II and Sanson's sinusoidal projection. The same holds for the distortion characteristics.

Transformation formulas

$$x = 1.89490\,R\lambda \left(\sqrt{1 - 3\frac{\phi^2}{\pi^2}} - \frac{1}{2} \right)$$

$$y = 0.94745\,R\phi$$

Distortion characteristics

$D_{ar} = 0.10$	$D_{an} = 30.8$	$D_{ab} = 0.39$
$D_{arc} = 0.11$	$D_{anc} = 31.4$	$D_{abc} = 0.40$

PSEUDOCYLINDRICAL PROJECTION WITH EQUALLY SPACED PARALLELS AND PARABOLIC MERIDIANS (PUTNINS)

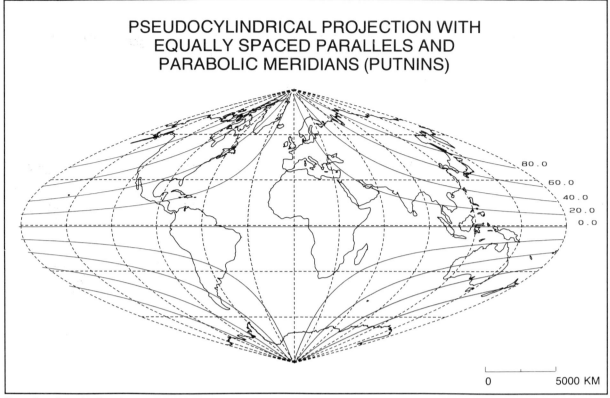

Lines of equal angular distortion

PSEUDOCYLINDRICAL PROJECTION WITH EQUALLY SPACED PARALLELS AND PARABOLIC MERIDIANS (PUTNINS)

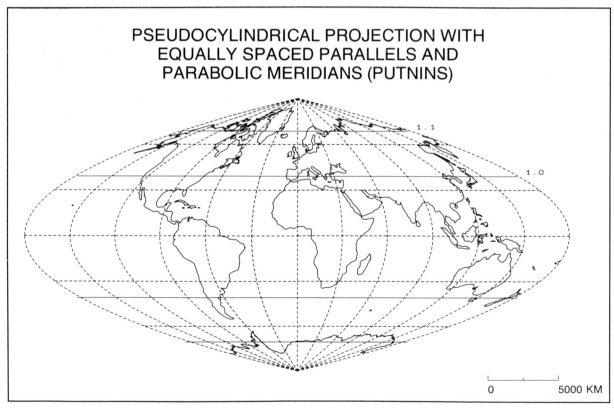

Lines of equal areal distortion

6.3 Pseudocylindrical projection with equally spaced parallels and parabolic meridians (Putnins)

Other name(s): Putnins P3 projection

Author(s): Reinholds V. Putnins (1934)

Description

The construction of Putnins P3 is very simple. The central meridian and equator are drawn as two perpendicular straight lines whereby the central meridian is half as long as the equator. The equator is then equally divided by parabolic meridians which pass through the poles. Finally the central meridian is equally divided by straight line parallels. The scale factor along the axes of the projection is chosen so that the total area of the map equals the area of the generating globe. This construction principle was also used to develop Arago's projection (the total area true version of Apianus II), Putnins P1 and Sanson's projection. The four projections differ from one another only in the steepness of the meridians and their graticules may be ordered according to this criterion. As is shown in Table 6.2 the nature of the meridians, which is the only variable in the construction of these projections, has a clear impact on the distortion characteristics. As the meridians converge sharper (full semiellipses ⇒ full cosinusoids) the angular distortion increases in favour of the areal distortion. The scale factor along the axes of the projection grows to compensate for the increasing steepness of the meridians so that the total area of the graticule remains equal to the area of the generating globe.

Table 6.2   Distortion characteristics of pseudocylindrical projections with equally spaced parallels

Projection	Nature of the meridians	Mean areal distortion	Mean angular distortion	Scale factor along the axes
Arago	Full ellipses	0.21	24.2	0.900
Putnins P1	Portions of ellipses	0.10	30.8	0.947
Putnins P3	Parabolas	0.04	35.4	0.977
Sanson	Full cosinusoids	0.00	39.0	1.000

Transformation formulas

$$x = \sqrt{\frac{3}{\pi}} R\lambda \left(1 - \frac{4\phi^2}{\pi^2} \right)$$

$$y = \sqrt{\frac{3}{\pi}} R\phi$$

Distortion characteristics

$D_{ar} = 0.04$ $D_{an} = 35.4$ $D_{ab} = 0.45$

$D_{arc} = 0.05$ $D_{anc} = 35.2$ $D_{abc} = 0.45$

SINUSOIDAL PROJECTION

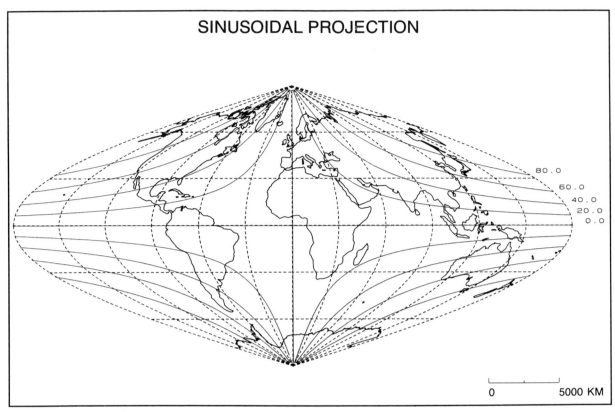

80.0
60.0
40.0
20.0
0.0

0   5000 KM

Lines of equal angular distortion

6.4   Pseudocylindrical equal-area projection with sinusoidal meridians (Sanson)

Other name(s):	Sanson–Flamsteed projection/Sinusoidal projection
Author(s):	Jean Cossin (1570)

Description

The projection is a limiting case of Bonne's pseudoconical equal-area projection with the equator taken as standard parallel. The parallels become equidistant straight lines while the meridians are represented by cosine curves.

The projection is less suited for world maps because of the considerable angular distortion near the edges of the map, especially in the higher latitudes. This results from the sinusoidal nature of the meridians. The mean angular distortion is relatively high for a pseudocylindrical equal-area projection.

The graticule of the projection is very easy to construct and although less important today this argument partly explains the historical importance of the projection. Sanson applied it for the first time in 1650. The 1606 edition of Mercator's *Atlas* on the other hand already contained a map of South America on this projection. Therefore the projection is in German literature often referred to as the 'Mercator–Sanson' projection. Keuning (1955) among others states that the graticule was developed in 1570 by Jean Cossin.

The projection is most frequently used for the mapping of continental areas with a pronounced N–S extension such as Africa and South America. It is also an interesting parent projection for the development of pseudocylindrical projections with sinusoidal meridians and pole line (Wagner, Eckert, McBryde and Thomas).

Transformation formulas

$$x = R\lambda \cos \phi$$

$$y = R\phi$$

Distortion characteristics

$D_{ar} = 0.00$	$D_{an} = 39.0$	$D_{ab} = 0.51$
$D_{arc} = 0.00$	$D_{anc} = 38.0$	$D_{abc} = 0.49$

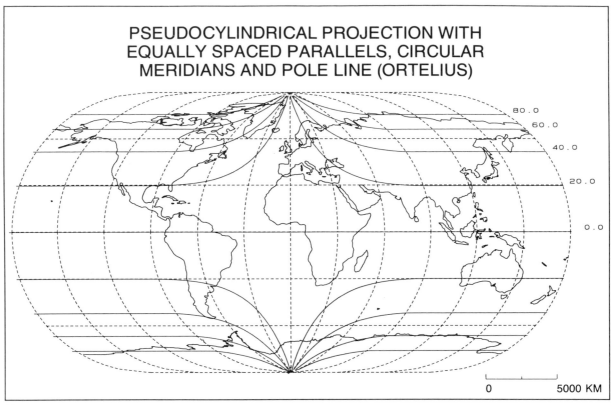

PSEUDOCYLINDRICAL PROJECTION WITH
EQUALLY SPACED PARALLELS, CIRCULAR
MERIDIANS AND POLE LINE (ORTELIUS)

Lines of equal angular distortion

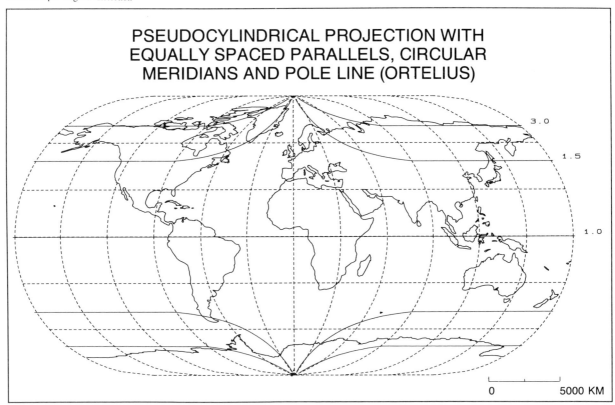

PSEUDOCYLINDRICAL PROJECTION WITH
EQUALLY SPACED PARALLELS, CIRCULAR
MERIDIANS AND POLE LINE (ORTELIUS)

Lines of equal areal distortion

6.5 Pseudocylindrical projection with equally spaced parallels, circular meridians and pole line

Author(s): Abraham Ortelius (1570)

Description

On this projection the inner hemisphere is represented by a closed circle. The meridians within this hemisphere are circular arcs which divide the equator in equal parts. The other meridians are half circles with the same radius as the inner hemisphere. They are equally spaced along the equidistantly extended equator. The result is a pseudocylindrical projection with a correct ratio of the axes and a pole line half the length of the equator.

This projection satisfies the definition of the so-called oval projection which was very popular in the sixteenth century but was seldom used afterwards. The typical oval projection is characterized by rectilinear parallels that divide the central meridian equidistantly, a 2/1 ratio of the axes and a pole line half the length of the equator. Two variants are described in the literature, one with circular and one with elliptical meridians. This projection corresponds to the first variant while Eckert's third projection (Eckert III) satisfies the definition of the second variant.

The oval projection may be distinguished from other historically important projections by:
(a) the use of a pole line;
(b) the distortion-compensating property.
It can be regarded as a precursor of the nowadays very popular projections in atlas cartography which give up the absolute equivalence to compensate the angular distortion (e.g. the Winkel–Tripel projection).

The pole-line concept is generally attributed to Petrus Apianus who gives two drafts of an oval projection in his *Cosmographicus liber* (1524). However, it was not until 1570 that a projection was published with a graticule that really satisfied the just given description of an oval projection. In that year Abraham Ortelius published his *Theatrum orbis terrarum* which is considered to be the first 'modern' atlas (Horn, 1961). The famous world map ('Typus orbis terrarum') is on an oval projection with circular meridians and has definitely contributed to the success of the atlas.

Transformation formulas

Depending on the value of λ the x- and y-coordinates of a point are obtained by the following expressions:

$$\lambda = 0$$

$$x = 0 \qquad\qquad y = R\phi$$

$$|\lambda| \leq \frac{\pi}{2}$$

$$x = x_0 \pm \sqrt{x_0^2 - y^2 + r^2} \qquad\qquad y = R\phi$$

with

$$x_0 = \frac{r}{\pi\lambda}\left(\lambda^2 - \frac{\pi^2}{4}\right)$$

$$|\lambda| > \frac{\pi}{2}$$

$$x = \frac{r}{\pi}\left(2\lambda \mp \pi \pm \sqrt{\pi^2 - \frac{y^2\pi^2}{r^2}}\right) \qquad y = R\phi$$

$$r = R\frac{\pi}{2}$$

where x_0 = centre of the meridional arc for the specified longitude λ, and r = the length of the semi minor axis of the planisphere.

The $\pm$ signs in the formulas are to be interpreted as follows: the upper sign holds for $\lambda \geq 0$, the lower sign for $\lambda < 0$.

Distortion characteristics

$D_{ar} = 0.39$	$D_{an} = 21.6$	$D_{ab} = 0.29$
$D_{arc} = 0.54$	$D_{anc} = 25.2$	$D_{abc} = 0.38$

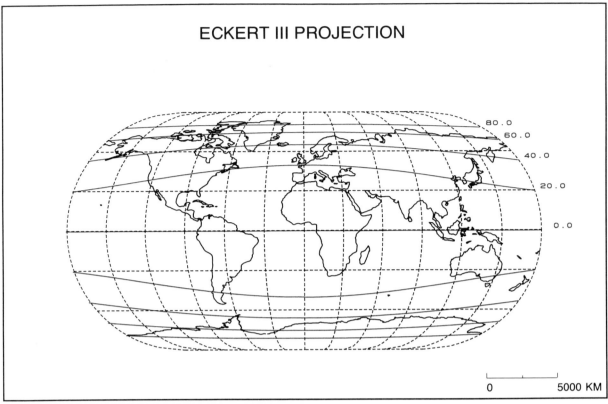

ECKERT III PROJECTION

80.0
60.0
40.0
20.0
0.0

0   5000 KM

Lines of equal angular distortion

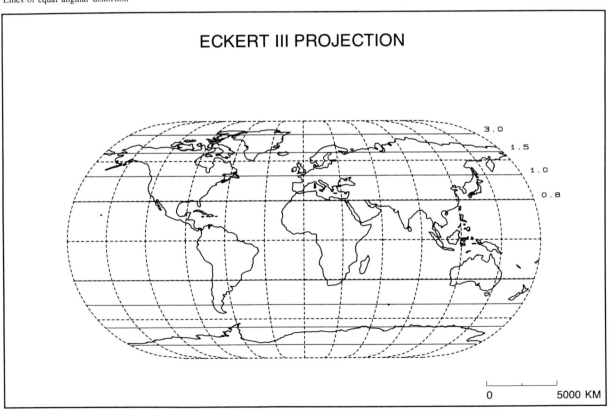

ECKERT III PROJECTION

3.0
1.5
1.0
0.8

0   5000 KM

Lines of equal areal distortion

6.6 Pseudocylindrical projection with equally spaced parallels, elliptical meridians and pole line (Eckert)

Other name(s): Eckert III projection

Author(s): Max Eckert (1906)

Description

M. Eckert defined his third projection as the arithmetic mean of Apianus' elliptical projection (Apianus II) and the cylindrical equidistant projection with one standard parallel (plate *carrée*). Like all Eckert's projections it has a pole line half the length of the equator and a 2/1 ratio of the axes. Eckert adjusted the length of the axes to make the projection 'total area true'. This adjustment is equivalent to a scale reduction and therefore has no effect on the appearance of the graticule. Just as for Arago's projection the area enclosed by two meridians is preserved.

Eckert's third projection satisfies the definition of an oval projection with elliptical meridians. The oval projections were very popular in the sixteenth century and are characterized by equally spaced rectilinear parallels, a 2/1 ratio of the axes and a pole line half the length of the equator. Two variants of the oval projection exist, one with elliptical meridians and one with circular meridians. In French cartographic literature Eckert's third projection is often referred to as 'projection d'Ortelius'. This denomination is not exact because Ortelius' projection is an oval projection with circular meridians (pseudocylindrical projection with equally spaced parallels, circular meridians and pole line).

In appearance Eckert's third projection is comparable with Wagner's sixth projection (Wagner VI). The only difference between the two is that on Eckert's projection the meridians are complete semiellipses while Wagner's projection uses only a fixed portion of a semiellipse to represent a meridian.

Transformation formulas

$$x = \frac{\lambda}{\pi}\left(r + \sqrt{r^2 - y^2}\right)$$

$$y = 2r\frac{\phi}{\pi}$$

$$r = R\sqrt{\frac{4\pi}{4 + \pi}}$$

where r = half the length of the minor axis.

Distortion characteristics

$D_{ar} = 0.36$	$D_{an} = 18.2$	$D_{ab} = 0.28$
$D_{arc} = 0.43$	$D_{anc} = 20.8$	$D_{abc} = 0.33$

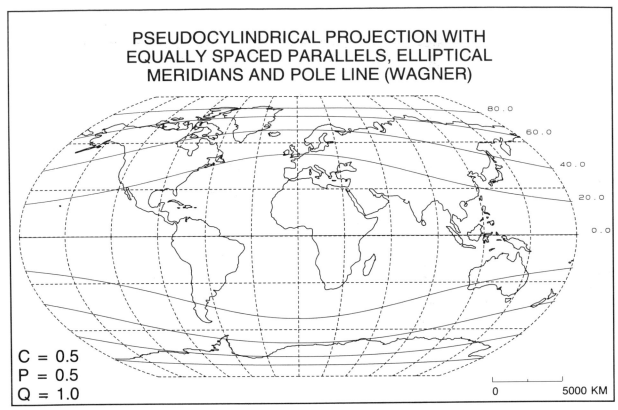

PSEUDOCYLINDRICAL PROJECTION WITH EQUALLY SPACED PARALLELS, ELLIPTICAL MERIDIANS AND POLE LINE (WAGNER)

C = 0.5
P = 0.5
Q = 1.0

0   5000 KM

Lines of equal angular distortion

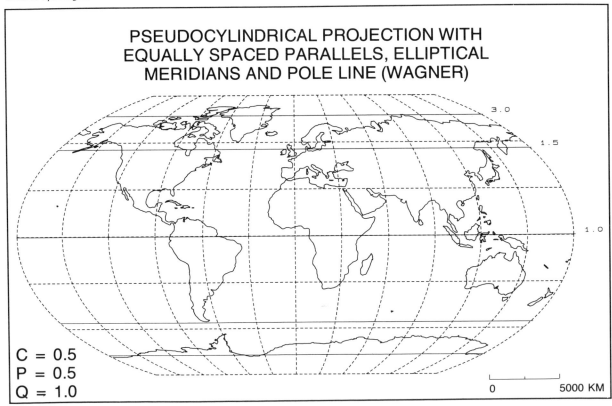

PSEUDOCYLINDRICAL PROJECTION WITH EQUALLY SPACED PARALLELS, ELLIPTICAL MERIDIANS AND POLE LINE (WAGNER)

C = 0.5
P = 0.5
Q = 1.0

0   5000 KM

Lines of equal areal distortion

6.7 Pseudocylindrical projection with equally spaced parallels, elliptical meridians and pole line (Wagner)

Other name(s): Wagner VI projection/Putnins P1′ projection

Author(s): Karlheinz Wagner (1962)/Reinholds V. Putnins (1934)

Description

Wagner developed this projection by transforming the graticule of Apianus' elliptical projection (Apianus II) through the introduction of parameters in the transformation formulas. By adjustment of these parameters it is possible to define projections with a given ratio of the axes (parameter p), length of the pole line (parameter c), and standard latitude (parameter q). Wagner chose $p = 0.5$ and $c = 0.5$. The result is a projection with a 2/1 ratio of the axes and a pole line half the length of the equator. By putting $q = 1.0$ the equator is represented in correct length. With $p = 0.5$, $c = 0.0$ and $q = 1.0$ the parent projection (Apianus II) is obtained.

Earlier Wagner transformed Sanson's sinusoidal projection in the same way (Wagner III). The resulting projection has similar distortion parameters. For Wagner's third projection (with sinusoidal meridians) the areal distortion is a little bit lower while his sixth projection has a lower angular distortion. Both belong to the group of projections which balance angular with areal distortion. Therefore they do not have the extreme distortions which appear on conformal and equal-area projections.

As Snyder (1977) pointed out, Putnins developed a projection which has the same graticule as Wagner's sixth projection with a pole line and a central meridian which are both half the length of the equator (Putnins P1′). The projection is the arithmetic mean of Putnins P1 and a plate *carrée* with the same central meridian and equator. The only difference between both projections is that Wagner maintained a true scale along the axes while Putnins reduced the scale of the graticule to make the total area of the map equal to the area of the generating globe.

Transformation formulas

$$x = Rq \frac{n\lambda}{\sqrt{nm}} \cos \psi \qquad\qquad y = R \frac{\pi}{2\sqrt{nm}} \sin \psi$$

$$\sin \psi = \frac{2m\phi}{\pi}$$

where $\qquad m = \sqrt{1 - c^2} \qquad\qquad n = \frac{m}{2p} \qquad q = \frac{\cos \phi_0}{\cos \psi_0}$

ϕ_0 = latitude of the standard parallel

and where $\qquad p = \dfrac{\text{central meridian}}{\text{equator}} \qquad\qquad c = \dfrac{\text{pole line}}{\text{equator}}$

The transformation formulas for Putnins P1′ may be derived from the general transformation formulas listed above. After simplification the following expressions are obtained:

$$x = 0.947\,45\, R\lambda \sqrt{1 - 3\frac{\phi^2}{\pi^2}} \qquad y = 0.947\,45\, R\phi$$

Distortion characteristics

Wagner VI	$D_{ar} = 0.33$	$D_{an} = 20.4$	$D_{ab} = 0.26$
	$D_{arc} = 0.46$	$D_{anc} = 22.2$	$D_{abc} = 0.31$
Putnins P1′	$D_{ar} = 0.27$	$D_{an} = 20.4$	$D_{ab} = 0.26$
	$D_{arc} = 0.37$	$D_{anc} = 22.2$	$D_{abc} = 0.30$

KAVRAISKY VII PROJECTION

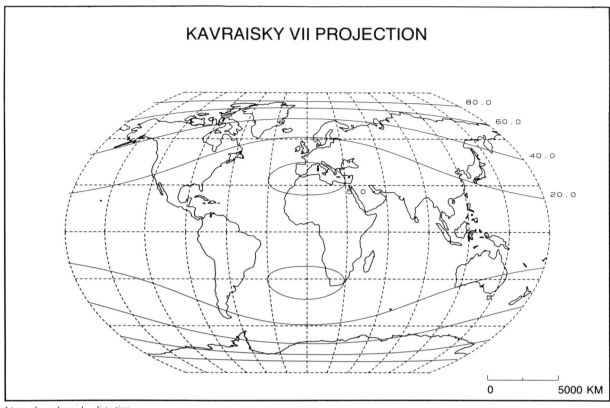

Lines of equal angular distortion

KAVRAISKY VII PROJECTION

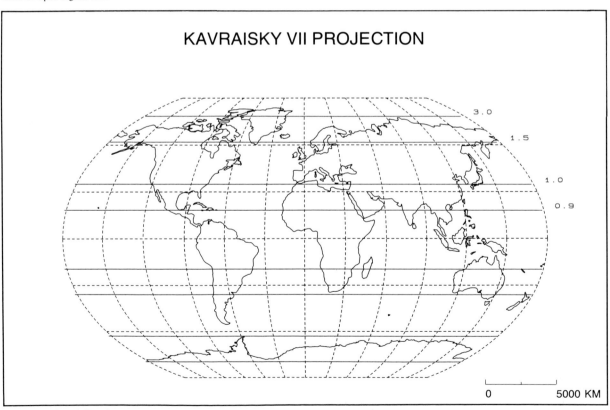

Lines of equal areal distortion

6.8 Pseudocylindrical projection with equally spaced parallels, elliptical meridians and pole line (Kavraisky)

Other name(s):   Kavraisky VII projection

Author(s):   V, V Kavraisky

Description

Kavraisky's seventh projection is very easy to construct. The central meridian is represented by a straight line of true length which is equally divided by the rectilinear parallels. The pole line is half the length of the equator and the meridians are arcs of ellipses which are equally spaced along the parallels. The meridians of 120° are arcs of a circle with its center in the origin of the projection. This leads to a incorrect ratio of the axes ($\sqrt{3}/1$ instead of $2/1$).

Compared with the other two projections with equally spaced parallels, elliptical meridians and pole line (Eckert III, Wagner VI), Kavraisky's seventh projection has a similar mean angular distortion although the distortion pattern is somewhat different (the minimum distortion occurs in the middle latitudes instead of on the equator). The areal distortion is considerably smaller than for the other projections. A comparison of the graticule of the three projections reveals a smaller E–W stretching in the higher latitudes on Kavraisky's projection as a result of the different ratio of the axes. This has a pleasing effect on the representation of the continents.

Transformation formulas

$$x = R\frac{\sqrt{3}}{2}\lambda \cos \psi$$

$$y = R\phi$$

$$\sin \psi = \frac{\sqrt{3}}{\pi}\phi$$

Distortion characteristics

$D_{ar} = 0.27$	$D_{an} = 19.1$	$D_{ab} = 0.23$
$D_{arc} = 0.36$	$D_{anc} = 20.6$	$D_{abc} = 0.26$

PSEUDOCYLINDRICAL PROJECTION WITH EQUALLY SPACED PARALLELS, PARABOLIC MERIDIANS AND POLE LINE (PUTNINS)

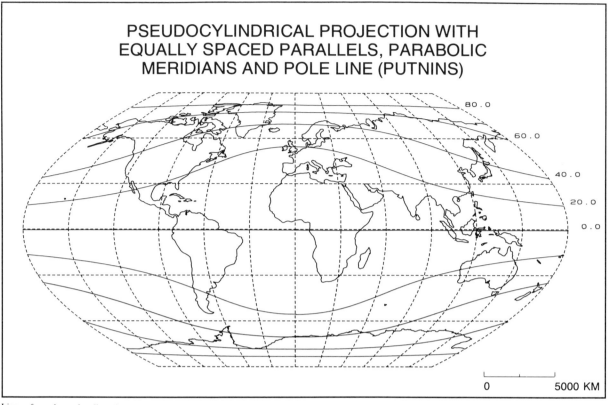

0   5000 KM

Lines of equal angular distortion

PSEUDOCYLINDRICAL PROJECTION WITH EQUALLY SPACED PARALLELS, PARABOLIC MERIDIANS AND POLE LINE (PUTNINS)

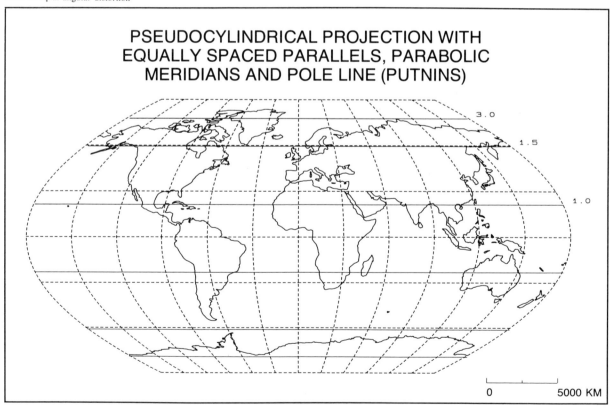

0   5000 KM

Lines of equal areal distortion

6.9 Pseudocylindrical projection with equally spaced parallels, parabolic meridians and pole line (Putnins)

Other name(s): Putnins P3' projection

Author(s): Reinholds V. Putnins (1934)

Description

All Putnins' pseudocylindrical projections have a correct ratio of the axes. For the projections with equally spaced parallels the scale along these axes remains constant but is less than 1.0 so the standard parallel does not coincide with the equator.

Putnins P3' is defined as the arithmetic mean of Putnins P3 and the cylindrical equidistant projection with one standard parallel (plate *carrée*). The latter is reduced so it has the same scale factor along the two axes as Putnins P3.

Since Putnins P3 has the same total area as the generating globe, Putnins P3' enlarges the total area as a result of the averaging with the plate *carrée*.

From the construction principle of the projection it follows that the pole line is half the length of the equator. The principle was also used by Putnins for the development of his two other projections with equally spaced parallels and pole line (P1' and P5') and by Eckert for the construction of his first, third and fifth projection (Eckert I, Eckert III, Eckert V). The graticule of Putnins P3' is very comparable with the Eckert V graticule although on the latter the meridians are somewhat sharper (full cosinusoids). The angular distortion patterns are similar for both projections. The areal distortion over continental area is less on Eckert's fifth projection because the parallel on which the areal scale factor equals 1.0 is better centered with respect to the large continents of the Northern Hemisphere.

Transformation formulas

$$x = \sqrt{\frac{3}{\pi}} R \lambda \left(1 - 2\frac{\phi^2}{\pi^2} \right)$$

$$y = \sqrt{\frac{3}{\pi}} R \phi$$

Distortion characteristics

$D_{ar} = 0.26$	$D_{an} = 22.1$	$D_{ab} = 0.27$
$D_{arc} = 0.37$	$D_{anc} = 23.2$	$D_{abc} = 0.31$

WINKEL II PROJECTION

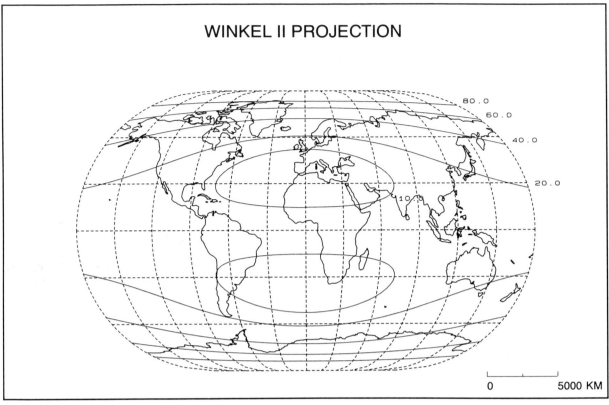

Lines of equal angular distortion

WINKEL II PROJECTION

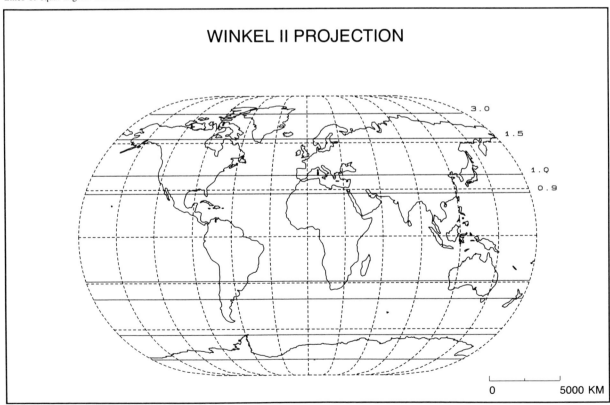

Lines of equal areal distortion

6.10 Pseudocylindrical projection with equally spaced parallels, elliptical meridians and pole line (Winkel)

Other name(s):	Winkel II projection
Author(s):	Oswald Winkel (1920)

Description

Winkel's second projection is the arithmetic mean of the cylindrical equidistant projection which has the same total area as the globe (standard parallels at 50° 28′ N and S latitude) and the pseudocylindrical projection with equally spaced parallels and elliptical meridians (Apianus II).

Just like on Winkel's two other graticules (Winkel I, Winkel–Tripel) the scale is true along the central meridian, the ratio of the axes is incorrect (the equator is $k + 1$ times the length of the central meridian) and the pole line is shorter than half the length of the equator (the pole line/equator ratio is $k/(k+1)$). Winkel's first and second projection differ from one another only in the nature of the meridians. While on Winkel I the meridians are sinusoids, on Winkel II they are complete semiellipses which are described by

$$\frac{(x - Rk(\lambda/2))^2}{R^2(\lambda^2/4)} + \frac{y^2}{R^2(\pi^2/4)} = 1$$

with $k = \cos \phi_0 = 0.6366$, and λ = longitude of the meridian.

In the lower latitudes the representation of the continents is similar on both projections. In the higher latitudes the longitudinal variation in angular distortion is less on Winkel's second graticule due to the nature of the meridians. This is reflected in lower angular distortion parameters and a more pleasing global representaion of the continents, resulting, however, in a larger areal exaggeration at higher latitudes.

From the mathematical description of the meridians it immediately follows that by putting $k = 1$ a projection is obtained with a correct ratio of the axes, a pole line half the length of the equator and outer meridians which are semicircles. This projection is the arithmetic mean of the plate *carrée* and Apianus II and is known as Eckert's third projection (Eckert III) (Eckert introduced a multiplier in the transformation formulas of the projection to make it total area true). Because of the correct ratio of the axes, Eckert's third projection gives a better representation of the continental areas in the lower latitudes than Winkel II which stretches the continents in a N–S direction. However, Eckert III is characterized by a visually disturbing E–W stretching in the higher latitudes. Accordingly the areal distortion parameters have substantially higher values than for Winkel's second projection.

Transformation formulas

$$x = \frac{R\lambda}{2}\left(k + \sqrt{1 - \frac{4\phi^2}{\pi^2}}\right)$$

$$y = R\phi$$

$$k = \cos \phi_0 = \frac{2}{\pi} = 0.6366$$

Distortion characteristics

$D_{ar} = 0.29$	$D_{an} = 18.4$	$D_{ab} = 0.22$
$D_{arc} = 0.38$	$D_{anc} = 20.1$	$D_{abc} = 0.25$

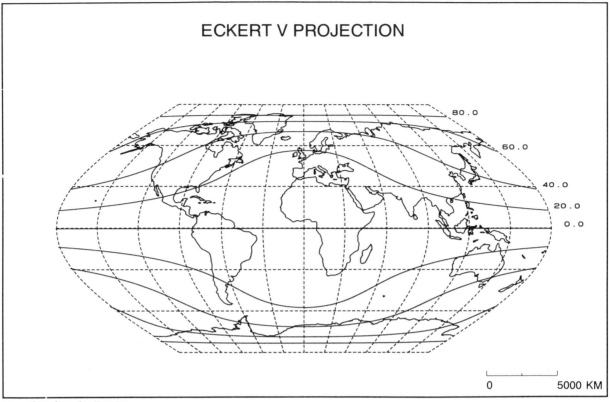

ECKERT V PROJECTION

80.0

60.0

40.0

20.0

0.0

0   5000 KM

Lines of equal angular distortion

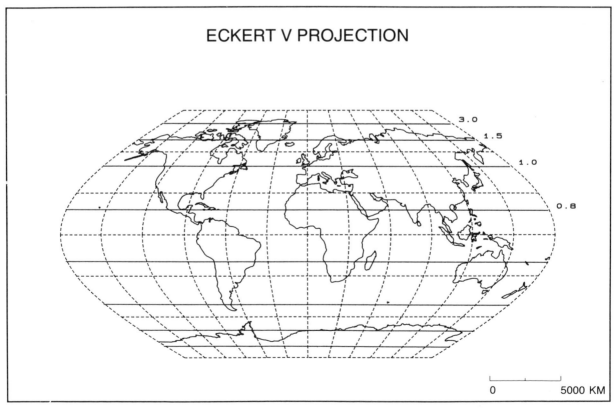

ECKERT V PROJECTION

3.0

1.5

1.0

0.8

0   5000 KM

Lines of equal areal distortion

6.11   Pseudocylindrical projection with equally spaced parallels, sinusoidal meridians and pole line (Eckert)

Other name(s):   Eckert V projection

Author(s):   May Eckert (1906)

Description

This projection is the arithmetic mean of Sanson's pseudocylindrical equal-area projection and the cylindrical equidistant projection with one standard parallel (plate *carrée*). Just as for Eckert's other projections the ratio of the axes is correct (2/1) and the pole line is half the length of the equator. The length of the axes is adjusted to make the projection 'total area true'. This has no effect on the appearance of the graticule and on the angular distortion. Since this adjustment is equivalent to a scale reduction only the areal distortion is changed.

Compared with Eckert's third projection the mean angular distortion is higher. This is a consequence of the type of curve used for the meridians. Sinusoidal meridians always give rise to higher angular distortions than elliptical meridians. The sinusoidal curve is steep and leads to a considerable compression near the edges of the map. On the other hand the areal distortion is smaller for Eckert's fifth projection.

The graticule of Eckert's fifth projection shows much resemblance to Wagner's third projection. The only difference is that the meridians on Eckert's fifth projection are full cosinusoïds while Wagner's third projection only uses a portion of the cosinusoid.

Transformation formulas

$$x = r\frac{\lambda}{\pi}(1 + \cos \phi)$$

$$y = 2r\frac{\phi}{\pi}$$

$$r = R\frac{\pi}{\sqrt{\pi + 2}}$$

where r = half the length of the minor axis.

Distortion characteristics

$D_{ar} = 0.28$	$D_{an} = 23.4$	$D_{ab} = 0.30$
$D_{arc} = 0.34$	$D_{anc} = 24.0$	$D_{abc} = 0.32$

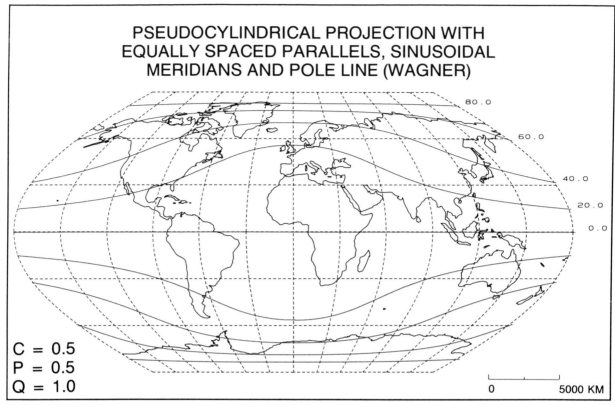

Lines of equal angular distortion

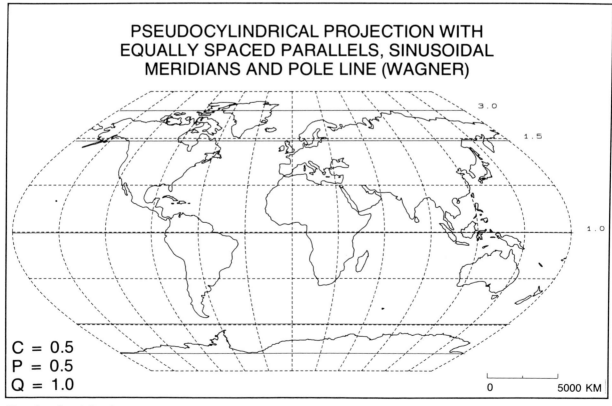

Lines of equal areal distortion

6.12 Pseudocylindrical projection with equally spaced parallels, sinusoidal meridians and pole line (Wagner)

Other name(s):   Wagner III projection

Author(s):   Karlheinz Wagner (1962)

Description

Wagner developed this projection by transforming the graticule of Sanson's sinusoidal projection. Three parameters p, c and q in the general transformation formulas of the projection allow for the adjustment of the ratio of the axes (p), the length of the pole line (c) and the latitude of the parallel which is to be represented in correct length (q). Wagner chose $p = 0.5$, $c = 0.5$ and $q = 1.0$.

A comparison with Eckert V, a projection with similar appearance, reveals that the distortion characteristics are comparable with exception of the areal distortion over continental area which is more substantial for Wagner's graticule. The distortion patterns explain this. The areal distortion on Eckert's fifth projection reaches a minimum in the middle latitudes while on Wagner's projection a minimum is reached on the equator.

Pseudocylindrical projections with equally spaced parallels were seldom used in the past because of their relatively high areal distortion. Nowadays most cartographers realize that when no absolute equivalence is required (e.g. for a general reference map) it is better to use a projection that is neither equal-area nor conformal. Projections of this kind have less extreme distortions and therefore give a more balanced representation of the earth's main features. The pseudocylindrical projections with equally spaced parallels belong to this group of projections.

Transformation formulas

$$x = Rq \frac{n\lambda}{\sqrt{nm}} \cos(m\phi) \qquad y = R \frac{m\phi}{\sqrt{nm}}$$

$$m = \frac{2 \arccos c}{\pi} \qquad n = \frac{m}{2p} \qquad q = \frac{\cos \phi_0}{\cos(2\phi_0/3)}$$

where ϕ_0 = latitude of the standard parallel, and

$$p = \frac{\text{central meridian}}{\text{equator}} \qquad c = \frac{\text{pole line}}{\text{equator}}$$

Distortion characteristics

$D_{ar} = 0.29$	$D_{an} = 22.6$	$D_{ab} = 0.28$
$D_{arc} = 0.40$	$D_{anc} = 23.5$	$D_{abc} = 0.31$

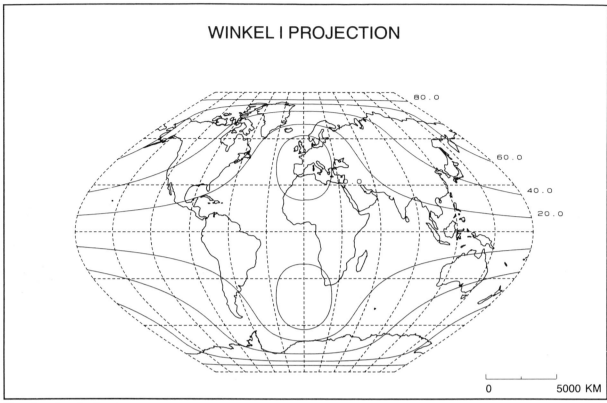

Lines of equal angular distortion

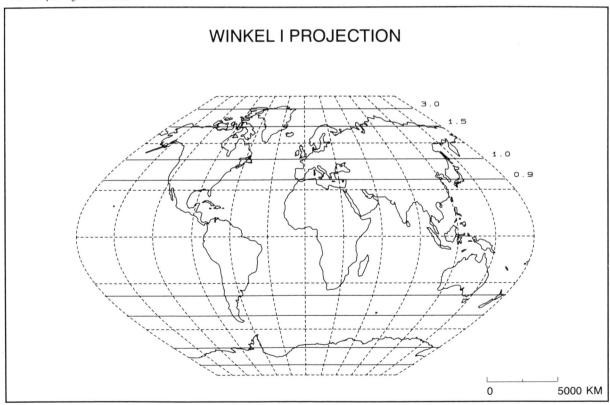

Lines of equal areal distortion

6.13 Pseudocylindrical projection with equally spaced parallels, sinusoidal meridians and pole line (Winkel)

Other name(s): Winkel I projection

Author(s): Oswald Winkel (1928)/Ronald Miller (1949)

Description

O. Winkel described three projections with equally spaced parallels which are the arithmetic mean of the cylindrical equidistant projection with two standard parallels and three different pointed-polar projections with true scale along the central meridian. He calculated the standard latitude of the cylindrical equidistant projection as follows:

$$k = \cos \phi_0 = \frac{\sin \phi_1}{\phi_1}$$

with ϕ_1 indicating the outer parallel of the zone that is to be represented. For the representation of the whole world ϕ_0 becomes $50° 28'$ which corresponds to a cylindrical equidistant projection that has the same total area as the generating globe.

Winkel's first projection is the arithmetic mean of the cylindrical equidistant projection and Sanson's pseudo-cylindrical equal-area projection with sinusoidal meridians. The projection is total area true and maintains true scale along the central meridian. Since for Sanson's projection scale is correct along every parallel, Winkel's first projection has the same standard parallel as the cylindrical equidistant projection which was used to construct it. So for a world map the standard parallels of Winkel I are situated at $50° 28'$ N and S latitude. All parallels at lower latitudes are reduced in length while those at higher latitudes are enlarged. The ratio of the axes is given by $k+1$. This means that for a world map the equator is $2/\pi + 1$ or 1.637 times larger than the central meridian. A correct ratio of the axes (2/1) is obtained by averaging Sanson's projection with the plate *carrée* (cylindrical equidistant projection with one standard parallel). This projection is known as Eckert's fifth (Eckert V). Since the plate *carrée* has not the same area as the generating globe Eckert introduced a multiplier (0.882)

in the transformation formulas of his projection to make it total area true.

As a result of the incorrect ratio of the axes the graticule of Winkel's first projection shows a relative elongation of the N–S distances in the lower latitudes. Eckert's fifth projection on the contrary has no angular distortion along the equator. Therefore it gives a more familiar representation of Africa and South America. Moreover the longitudinal variation of the angular distortion is less on Eckert's projection. On Winkel I it increases at a considerable rate toward the edges of the map, e.g. the representation of Europe is virtually conformal while North America has an angular distortion which varied between $20°$ and $60°$. This longitudinal variation of angular distortion is typical for projections with sinusoidal meridians but is more pronounced as the length of the pole line decreases (it is extreme on Sanson's graticule).

It must be pointed out that R. Miller developed independently a projection identical to Winkel's first. Miller is credited with the cylindrical equidistant projection which has the same total area as the globe. In 1949 he proposed a projection which is the arithmetic mean of this so-called equi-rectangular projection and Sansons's sinusoidal.

Transformation formulas

$$x = \frac{R\lambda}{2}(k + \cos \phi)$$

$$y = R\phi$$

$$k = \cos \phi_0 = \frac{2}{\pi} = 0.6366$$

Distortion characteristics

$D_{ar} = 0.22$	$D_{an} = 25.8$	$D_{ab} = 0.29$
$D_{arc} = 0.27$	$D_{anc} = 25.6$	$D_{abc} = 0.29$

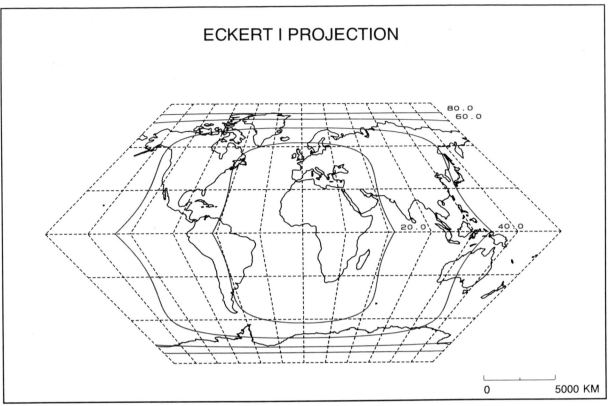

ECKERT I PROJECTION

0   5000 KM

Lines of equal angular distortion

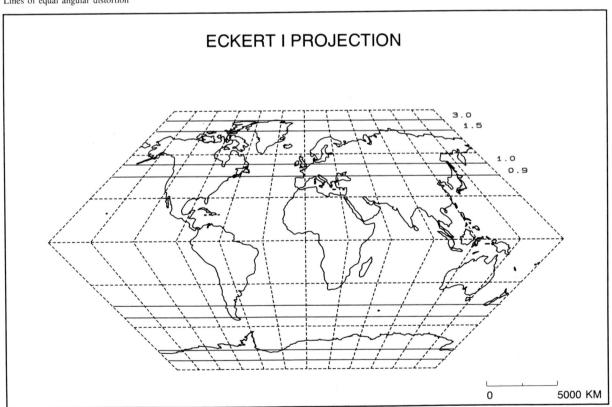

ECKERT I PROJECTION

0   5000 KM

Lines of equal areal distortion

6.14 Pseudocylindrical projection with equally spaced parallels, rectilinear meridians and pole line (Eckert)

Other name(s): Eckert I projection

Author(s): Max Eckert (1906)

Description

In 1906 M. Eckert defined six pseudocylindrical projections for world maps which are all characterized by a correct ratio of the axes and a pole line half the length of the equator. The first two projections have rectilinear meridians, the third and the fourth have elliptical meridians. The last two projections are characterized by sinusoidal meridians. The first, third and fifth projection (Eckert I, Eckert III, Eckert V) have equally spaced parallels. For the other three projections (Eckert II, Eckert IV, Eckert VI) the spacing of the parallels decreases with increasing latitude. These projections are equal-area.

The projections with equally spaced parallels are all defined as the arithmetic mean of a pseudocylindrical projection which shows the pole as a point and the cylindrical equidistant projection with one standard parallel (plate *carrée*). The length of the axes is adjusted to make the total area of the planisphere equal to the area of the globe. The other three projections are equal-area versions of these.

Eckert's first projection is the simplest of all but it has no real practical value. Since the meridians are straight lines which are equally spaced along the rectilinear parallels they meet at an angle on the equator. This discontinuity leads to a very unpleasing appearance of the equatorial areas.

Transformation formulas

$$x = \frac{\lambda}{\pi}(2r - |y|)$$

$$y = 2r\frac{\phi}{\pi}$$

$$r = R\sqrt{\frac{2}{3}\pi}$$

where r = half the length of the minor axis.

Distortion characteristics

$D_{ar} = 0.28$	$D_{an} = 30.3$	$D_{ab} = 0.35$
$D_{arc} = 0.34$	$D_{anc} = 26.4$	$D_{abc} = 0.32$

MOLLWEIDE PROJECTION

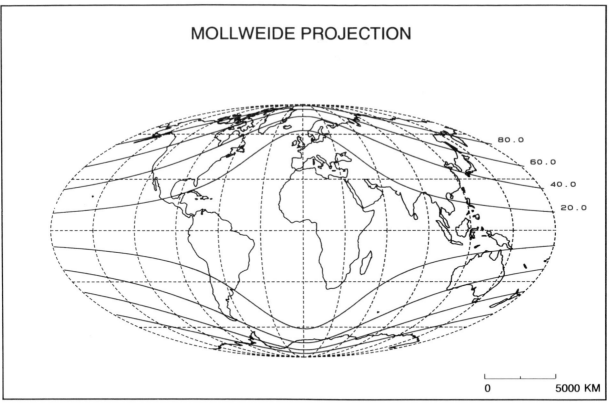

80.0
60.0
40.0
20.0

0   5000 KM

Lines of equal angular distortion

TRANSVERSE ASPECT MOLLWEIDE
PROJECTION WITH LONGER AXIS 25W, 155E

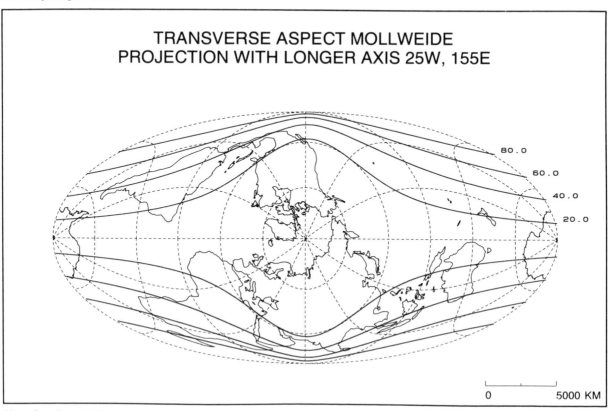

80.0
60.0
40.0
20.0

0   5000 KM

Lines of equal angular distortion

6.15 Pseudocylindrical equal-area projection with elliptical meridians (Mollweide)

6.15.1 Normal Aspect

Other name(s): Homolographic projection

Author(s): Karl Mollweide (1805)

Description

Mollweide's projection is the equivalent version of Apianus' second projection (Apianus II). This implies that the total area of the bounding ellipse equals the area of the generating globe. Moreover, the distance between the parallels decreases with increasing latitude to ensure the equal-area property.

The projection was developed in 1805 by the German mathematician Karl Mollweide, but it was not until the French cartographer Jacques Babinet (1857) popularized it as the 'projection homolographique' that it became well known. Since then the projection has been very popular in atlas cartography for world maps showing distributions.

Transformation formulas

$$x = 2R\sqrt{2}\,\frac{\lambda}{\pi}\,\cos\psi \qquad\qquad y = R\sqrt{2}\,\sin\psi$$

$$2\psi + \sin 2\psi = \pi\,\sin\phi$$

The relation between the geographical latitude ϕ and the auxiliary angle ψ is expressed by a transcendental equation which must be solved numerically.

Distortion characteristics

$D_{ar} = 0.00$	$D_{an} = 32.3$	$D_{ab} = 0.39$
$D_{arc} = 0.00$	$D_{anc} = 33.7$	$D_{abc} = 0.42$

As the whole world is represented within an ellipse no discontinuity appears at the poles. Therefore Mollweide's projection lends itself very well to the development of transverse and oblique aspects. A change of aspect does not alter the distortion pattern of the projection but only repositions it with respect to the surface of the earth. Therefore the values of the global distortion parameters equal the values for the normal aspect of the projection. Only the distortion parameters over continental area will change. The transformation formulas for a transverse or oblique aspect may be derived from the formulas of the normal aspect by applying the rules of spherical trigonometry (Section 2.2).

6.15.2 Transverse Aspect

Description

As described in Section 2.2, a transverse aspect of a pseudocylindrical projection may be generated in two ways. The distortion pattern may be rotated without moving its center or the center may be moved to coincide with one of the geographical poles. In the latter case it is possible to optimize the choice of the axes of the projection by minimizing an overall distortion parameter. When the North Pole is taken as the centre of projection the minimization of D_{abc} leads to a transverse aspect Mollweide projection with as the greater axis the meridians 25° W, 155° E (rounded off to the nearest 5°). This aspect shows very well the relative positions of the continental areas in the Northern Hemisphere. The choice of the greater axis at 25° W, 155° E guarantees that except from Antarctica no continents are interrupted.

A comparison of the distortion parameters with those of the normal aspect Mollweide projection illustrates that a change of aspect allows us to reduce the scale distortion over the continental area by a factor of two while the angular distortion is also considerably lowered.

Distortion characteristics

$D_{arc} = 0.00$	$D_{anc} = 22.0$	$D_{abc} = 0.23$

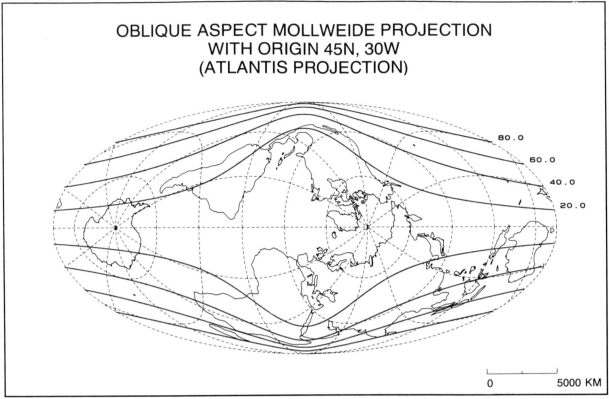

OBLIQUE ASPECT MOLLWEIDE PROJECTION
WITH ORIGIN 45N, 30W
(ATLANTIS PROJECTION)

80.0
60.0
40.0
20.0

0   5000 KM

Lines of equal angular distortion

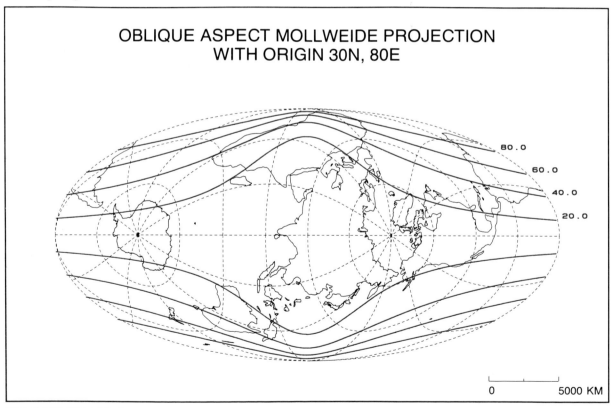

OBLIQUE ASPECT MOLLWEIDE PROJECTION
WITH ORIGIN 30N, 80E

80.0
60.0
40.0
20.0

0   5000 KM

Lines of equal angular distortion

6.15 Pseudocylindrical equal-area projection with elliptical meridians (Mollweide) (CONTINUED)

6.15.3 Oblique Aspect

Oblique aspect with a central meridian as greater axis

1. Atlantis projection

Author(s): John Bartholomew

Description

One of the best known oblique aspect Mollweide projections is the so-called Atlantis projection which was developed by John Bartholomew. The greater axis of this projection is formed by the meridians of 30° W and 150° E. The smaller axis is a great circle that reaches its vertex at 45° N, 30° W. This point is the centre of the projection.

The Atlantis projection gives a much less distorted representation of the continents than the normal aspect Mollweide projection (compare the distortion parameters). With the exception of New Zealand no continental areas are interrupted. Both polar areas are represented well.

Distortion characteristics $\qquad D_{arc} = 0.00 \qquad D_{anc} = 22.0 \qquad D_{abc} = 0.23$

2. Minimum-error projections

Description

Minimization of the scale distortion over the continental area (D_{abc}) leads to a centre of projection at 45° N, 35° W. This position approximates closely the centre of the Atlantis projection. Hence the distortion parameters of both aspects do not differ much.

Apart from this minimum-error projection, the minimization of D_{abc} for the oblique aspect Mollweide projection with a central meridian as the greater axis gives a second local

minimum for a centre of projection situated at 30° N, 80° E. The distortion parameters have slightly higher values than for the first solution. In contrast to the Atlantis projection this second minimum-error projection places the Indian Ocean in the centre. Therefore the Eastern Hemisphere is represented better. A disadvantage of this aspect is the interruption of South America.

Distortion characteristics

The minimum-error projection (center at 45° N, 35° W) has the following distortion parameters.

$$D_{arc} = 0.00 \qquad D_{anc} = 20.4 \qquad D_{abc} = 0.21$$

For the second solution (center at 30° N, 80° E) slightly higher values are obtained:

$$D_{arc} = 0.00 \qquad D_{anc} = 21.6 \qquad D_{abc} = 0.23$$

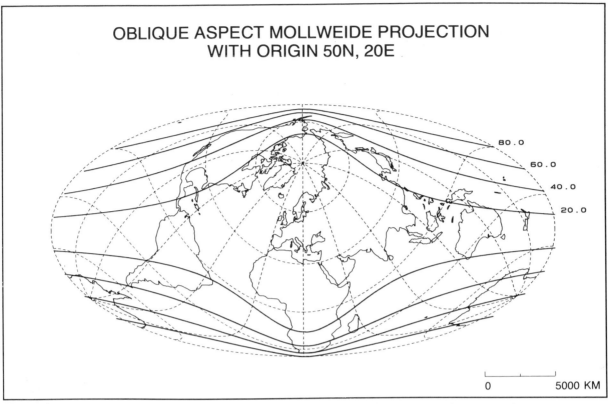

OBLIQUE ASPECT MOLLWEIDE PROJECTION
WITH ORIGIN 50N, 20E

80.0
60.0
40.0
20.0

0   5000 KM

Lines of equal angular distortion

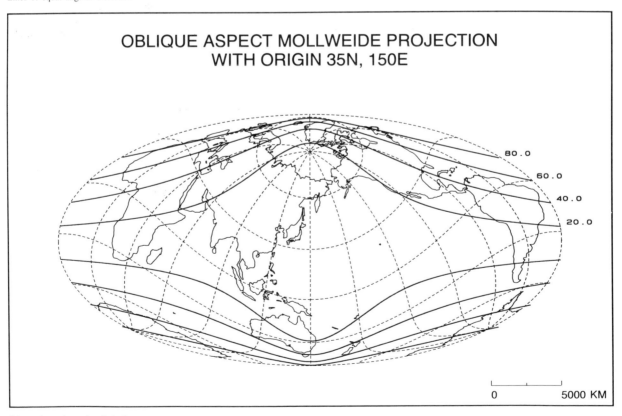

OBLIQUE ASPECT MOLLWEIDE PROJECTION
WITH ORIGIN 35N, 150E

80.0
60.0
40.0
20.0

0   5000 KM

Lines of equal angular distortion

6.15 Pseudocylindrical equal-area projection with elliptical meridians (Mollweide) (CONTINUED)

6.15.3 Oblique Aspect

Oblique aspect with a central meridian as smaller axis

It is also possible to develop oblique aspects with a central meridian as the smaller axis. The greater axis then becomes a great circle that reaches its vertex at the intersection with this central meridian.

1. Origin at 50° N, 20° E

Description

Minimization of the scale distortion over the continental area leads to a centre of projection at 55° N, 20° E. The resulting projection does not give a satisfactory representation of the African continent. This illustrates that the minimization of distortion cannot be handled as a single criterion in the selection of a suitable projection for a world map. Nevertheless the minimization procedure may suggest alternative representations that broaden the spectrum of possibilities and therefore make the final choice easier and more objective. In this case a translation of the centre of projection in the southern direction (50° N, 20° E) still gives a good representation of the continental area as a whole while Africa will be less distorted. The effect of this translation on the distortion parameters is minimal.

Distortion characteristics

For the minimum-error projection (centre at 55° N, 20° E) the distortion parameters are

$$D_{arc} = 0.00 \qquad D_{anc} = 19.4 \qquad D_{abc} = 0.20$$

A small translation of the centre of projection (50° N, 20° E) gives

$$D_{arc} = 0.00 \qquad D_{anc} = 19.9 \qquad D_{abc} = 0.21$$

2. Origin at 35° N, 150° E

Description

Due to the asymmetric distribution of continents and oceans over the globe, minimization over the continental surface generally leads to a graticule with the central meridian situated over Central Europe and Africa. However interesting views of the globe might be obtained by focusing on selected parts of the earth. By choosing the origin at 35° N, 150° E an oblique aspect Mollweide projection is obtained which does not intersect the major continents and which centers on eastern Asia and the western Pacific.

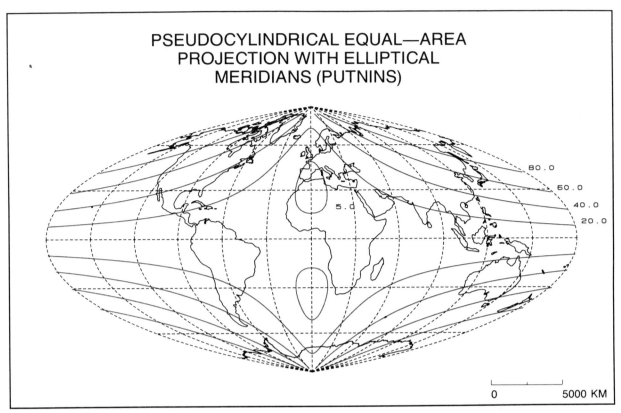

PSEUDOCYLINDRICAL EQUAL—AREA
PROJECTION WITH ELLIPTICAL
MERIDIANS (PUTNINS)

Lines of equal angular distortion

6.16 Pseudocylindrical equal-area projection with elliptical meridians (Putnins)

Other name(s): Putnins P2 projection

Author(s): Reinholds V. Putnins (1934)

Description

As explained in the description of Putnins's P1 projection this author proposed twelve pseudocylindrical projections with conic section meridians (elliptical, parabolic and hyperbolic meridians). Putnins P2 is the equal-area version of Putnins P1 with the distance between the parallels decreasing toward the poles to ensure the equal-area property.

Putnins P2 and Mollweide's projection are therefore of the same type, only Mollweide used full ellipses for the meridians while Putnins used a portion of a semiellipse for each meridian. Hence Putnins's graticule shows a discontinuity at the poles.

Putnins P2 is intermediate between Mollweide's and Sanson's equal-area projections. The scale variation along the central meridian is less than on Mollweide's graticule so Putnins's projection does not excessively elongate the N–S distances in the lower latitudes. On the other hand it avoids the high-latitude crowding which is characteristic of Sanson's projection as a consequence of the steep sinusoidal meridians.

Boggs's pseudocylindrical equal-area projection (Eumorphic projection), which is also intermediate between Mollweide and Sanson, has a graticule almost identical with Putnins P2 although the mathematical definition of both projections is completely different.

Transformation formulas

$$x = 1.894\,90 R\lambda \left(\cos \psi - \frac{1}{2} \right)$$

$$y = 1.718\,48 R \sin \psi$$

$$2\psi + \sin 2\psi - 2 \sin \psi = n \sin \phi$$

where $n = \dfrac{4\pi - 3\sqrt{3}}{6}$

The value of ψ which corresponds to a given value of ϕ may be calculated by numerical approximation.

Distortion characteristics

$D_{ar} = 0.00$	$D_{an} = 34.9$	$D_{ab} = 0.50$
$D_{arc} = 0.00$	$D_{anc} = 35.5$	$D_{abc} = 0.55$

BOGGS' EUMORPHIC PROJECTION

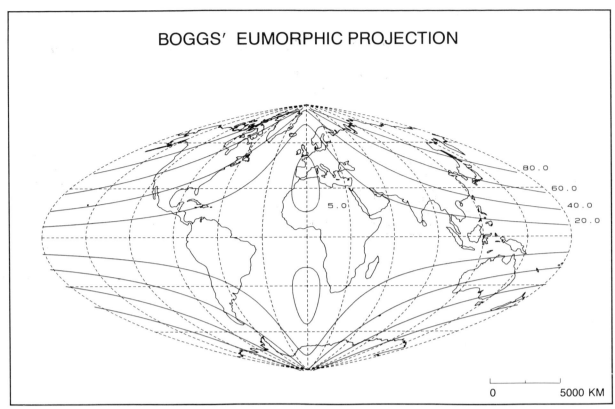

80.0
60.0
40.0
20.0

5.0

0   5000 KM

Lines of equal angular distortion

6.17 Pseudocylindrical equal-area projection (Boggs)

Other name(s):	Eumorphic projection
Author(s):	Whittemore Boggs (1929)

Description

In 1929 Boggs constructed this projection by averaging the y-coordinates of Sanson's and Mollweide's pseudocylindrical equal-area projections. To maintain the equal-area property of the original graticules the x-coordinate of the projection was derived from the general transformation formulas of the pseudocylindrical equal-area projection with equally divided parallels:

$$x = R^2 \lambda \, \frac{\cos \phi}{dy/d\phi}$$

$$y = Rf(\phi)$$

The principle of averaging two projections, thereby maintaining the equal-area property of the parent graticules was also used by other authors to construct new projection systems (Nell–Hammer).

Since Boggs wanted the ratio of the axes to be correct he multiplied the x-coordinate by a factor $k = 1.00138$ and divided the y-coordinate by the same value. This transformation does not affect the equal-area property.

With the term 'Eumorphism' Boggs refers to a good representation of large shapes, a property which according to him is obtained 'by approximation to rectilinear intersection of parallels and meridians and by the approximation to equality of linear scales along the parallels and meridians—or along all parallels and on at least a central meridian'.

Boggs's eumorphic projection is characterized by a mean angular distortion which is much lower than for Sanson's sinusoidal projection. Moreover, the projection does not show the excessive N–S compression in the higher latitudes characteristic of Mollweide's projection and resulting from the considerable scale variation along the central meridian. Therefore the projection may be seen as a good compromise between the two parent projections whereby the equal-area property is maintained.

The graticule of Boggs's projection is almost identical with the graticule of Putnins P2 although the mathematical description of both projections is completely different.

Transformation formulas

$$x = \frac{2Rk\lambda}{\sec \phi + 1.11072 \, \sec \psi}$$

$$y = \frac{1}{2k} R (\phi + \sqrt{2} \, \sin \psi)$$

$$\sin 2\psi + 2\psi = \pi \sin \phi$$

$$k = 1.00138$$

The transcendental expression which relates the auxiliary angle ψ to the geographical latitude ϕ is identical to the corresponding expression for Mollweide's projection.

Distortion characteristics			
	$D_{ar} = 0.00$	$D_{an} = 35.0$	$D_{ab} = 0.44$
	$D_{arc} = 0.00$	$D_{anc} = 35.1$	$D_{abc} = 0.44$

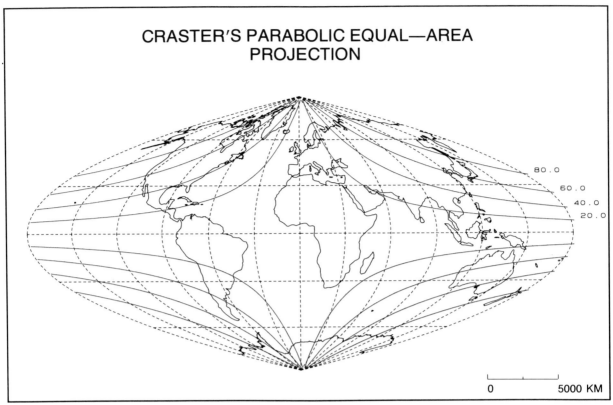

CRASTER'S PARABOLIC EQUAL—AREA PROJECTION

80.0
60.0
40.0
20.0

0   5000 KM

Lines of equal angular distortion

6.18 Pseudocylindrical equal-area projection with parabolic meridians (Craster)

Other name(s):	Craster's parabolic equal-area projection/ Putnins P4 projection
Author(s):	J. E. E. Craster (1929)/Reinholds V. Putnins (1934)

Description

It is possible to develop an infinite number of pseudocylindrical equal-area projections, intermediate between Sanson's and Mollweide's graticules and characterized by straight parallels which are equally divided by the meridians. Craster developed a series of conic section pseudocylindrical equal-area projections by gradually decreasing the eccentricity of the outer meridian. Mollweide's projection closes the series. Above all, Craster preferred the parabolic projection. It shows a considerable E–W compression in the higher latitudes. This is a common property of all pseudocylindrical equal-area projections that represent the pole as a point. Although the standard parallel is situated in the middle latitudes

the projection is virtually conformal along the equator (the maximum angular distortion is less than 3°). A comparison of the distortion parameters reflects the intermediate position of the parabolic projection between Sanson's and Mollweide's graticules.

In 1934 R. V. Putnins developed independently an identical projection (Putnins P4), which is the equal-area version of his pseudocylindrical projection with equally spaced parallels and parabolic meridians [Putnins P3]. The only difference between the graticules of Putnins P3 and Putnins P4 is that on the latter the spacing of the parallels decreases toward the poles to ensure the equal-area property.

Transformation formulas

$$x = 2r \left(2 \cos \frac{2\phi}{3} - 1 \right) \frac{\lambda}{\pi}$$

$$y = 2r \sin \frac{\phi}{3}$$

$$r = R \frac{\sqrt{3\pi}}{2}$$

where r = the length of the semi minor axis of the planisphere.

Distortion characteristics

$D_{ar} = 0.00$	$D_{an} = 37.0$	$D_{ab} = 0.48$
$D_{arc} = 0.00$	$D_{anc} = 36.6$	$D_{abc} = 0.47$

ADAMS' PSEUDOCYLINDRICAL EQUAL—AREA PROJECTION

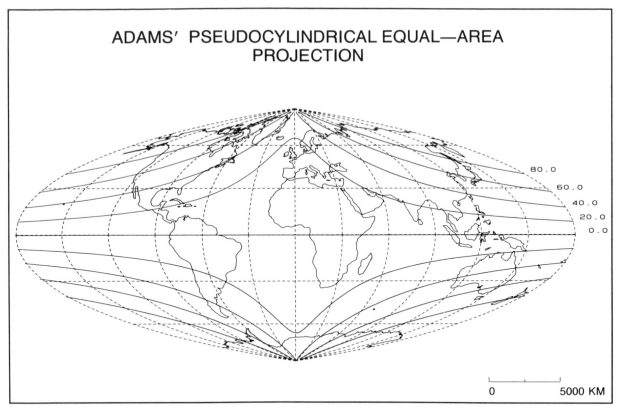

80.0
60.0
40.0
20.0
0.0

0   5000 KM

Lines of equal angular distortion

6.19 Pseudocylindrical equal-area projection with quartic meridians (Adams)

Other name(s): Adams's orthembadic projection

Author(s): Oscar S. Adams

Description

E. J. Baar (1947) described two subgroups of the pseudocylindrical equal-area projections, the so-called sine and tangent series. The sine series is characterized by the following general transformation formulas:

$$x = R\frac{q}{p} \lambda \cos \phi \sec \frac{\phi}{q}$$

$$y = Rp \sin \frac{\phi}{q}$$

The parameters p and q allow for the development of an unlimited number of projections that satisfy these general equations. Adams selected $p = q = 2$ to obtain his so-called 'orthembadic' (= equal-area) projection. The meridians in this projection are curves of fourth degree.

Just like Craster's pseudocylindrical equal-area projection with parabolic meridians (which also forms part of the tangent series) this projection is situated between Sanson's and Mollweide's pseudocylindrical equal-area projections with respect to the appearance of the graticule and the distortion characteristics.

The projection underlies the well-known pseudocylindrical equal-area projection with quartic meridians and pole line of McBryde and Thomas (flat-polar quartic projection).

Transformation formulas

$$x = R\lambda \cos \phi \sec \frac{\phi}{2}$$

$$y = 2R \sin \frac{\phi}{2}$$

Distortion characteristics

$D_{ar} = 0.00$	$D_{an} = 36.0$	$D_{ab} = 0.47$
$D_{arc} = 0.00$	$D_{anc} = 36.4$	$D_{abc} = 0.48$

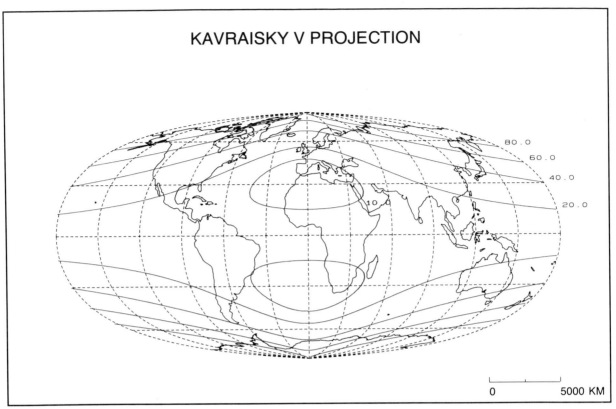

KAVRAISKY V PROJECTION

80.0
60.0
40.0
20.0
10.0

0   5000 KM

Lines of equal angular distortion

6.20   Pseudocylindrical equal-area projection (Kavraisky)

Other name(s):   Kavraisky V projection

Author(s):   V. V. Kavraisky (1933)

Description

In 1933 the Russian cartographer V. V. Kavraisky developed a pseudocylindrical equal-area projection which was used for a map of the Pacific in a Soviet atlas, published in 1940.

It is known as Kavraisky's fifth projection. In his review of Russian map projections Maling (1960) gives the following transformation formulas for the projection

$$x = R\frac{1}{ab}\lambda \; \sec(b\phi) \; \cos\phi \qquad y = Ra \; \sin(b\phi)$$

with $a = 1.504\,875$ and $b = 0.738\,341$.

Snyder (1977) pointed out that these transformation formulas fit the sine series, described by E. J. Baar in 1947 and used by McBryde and Thomas (1949) for the development of their so-called flat-polar authalic projections. The sine series is defined as

$$x = R\frac{q}{p}\lambda \; \cos\phi \; \sec\frac{\phi}{q} \qquad y = Rp \; \sin\frac{\phi}{q}$$

With $p = a$ and $q = 1/b$ Kavraisky's fifth projection is obtained. The values of the constants p and q correspond to a standard parallel at 35° N and S latitude while the scale along the equator is 0.9 of true scale (Snyder, 1977).

Transformation formulas

$$x = R\frac{q}{p}\lambda \; \cos\phi \; \sec\frac{\phi}{q} \qquad y = Rp \; \sin\frac{\phi}{q}$$

with $p = 1.504\,875$, and $q = 1.354\,388$.

Distortion characteristics

$D_{ar} = 0.00$	$D_{an} = 30.5$	$D_{ab} = 0.38$
$D_{arc} = 0.00$	$D_{anc} = 32.9$	$D_{abc} = 0.43$

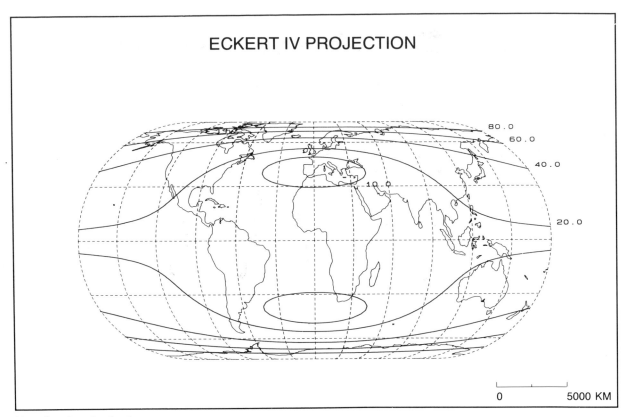

ECKERT IV PROJECTION

Lines of equal angular distortion

6.21 Pseudocylindrical equal-area projection with elliptical meridians and pole line (Eckert)

Other name(s): Eckert IV projection

Author(s): Max Eckert (1906)

Description

This projection is the equivalent version of Eckert's pseudocylindrical projection with equally spaced parallels, elliptical meridians and pole line (Eckert III). Both projections are similar in appearance. The only difference is that in Eckert's fourth projection the distance between the parallels decreases with increasing latitude. This is a necessary condition to ensure the equal-area property.

Behrmann's study (see Part I, p.34) indicated that Eckert's fourth projection has a very low mean angular distortion for an equal-area projection. Nevertheless the projection shows a considerable elongation of the N–S distances in the lower latitudes. Therefore it is less suited for world maps than certain equal-area projections with a higher mean angular distortion. As already explained in Section 3.1, the minimization of a specific distortion parameter does not lead to the 'best' projection for a particular purpose. Other factors are involved in the selection process that are not directly related to the distortion characteristics of the projection. Eckert preferred the pseudocylindrical equal-area projection with sinusoidal meridians (Eckert VI) above this one.

Transformation formulas

$$x = r\frac{\lambda}{\pi}(1 + \cos \psi)$$

$$y = r \sin \psi$$

$$2\psi + 4 \sin \psi + \sin 2\psi = (4 + \pi) \sin \phi$$

$$r = R\sqrt{\frac{4\pi}{4 + \pi}}$$

where r = the length of the semi minor axis of the planisphere.

The relation between the geographical latitude ϕ and the auxiliary angle ψ is expressed by a transcendental equation. The value of ψ that corresponds to a specific value of ϕ has to be calculated by numerical approximation.

Distortion characteristics

$D_{ar} = 0.0$	$D_{an} = 28.7$	$D_{ab} = 0.35$
$D_{arc} = 0.0$	$D_{anc} = 31.4$	$D_{abc} = 0.43$

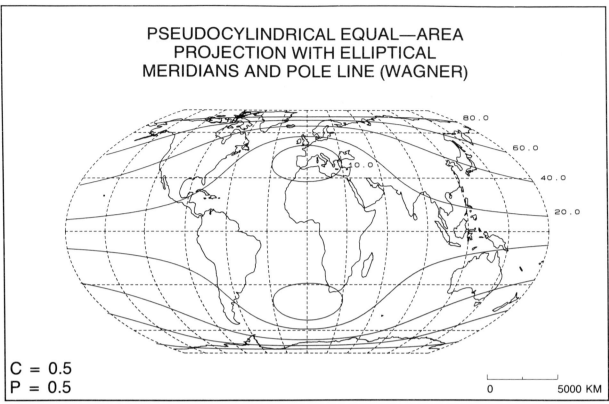

PSEUDOCYLINDRICAL EQUAL—AREA
PROJECTION WITH ELLIPTICAL
MERIDIANS AND POLE LINE (WAGNER)

C = 0.5
P = 0.5

0   5000 KM

Lines of equal angular distortion

6.22 Pseudocylindrical equal-area projection with elliptical meridians and pole line (Wagner)

Other name(s):	Wagner IV projection/Putnins P2′ projection/ Werenskiold III projection
Author(s):	Karlheinz Wagner (1962)/R. V. Putnins (1934)/ W. Werenskiold (1944)

Description

Wagner developed a general theory to transform existing projections through the introduction of parameters in the transformation formulas (Section 2.4). By putting some constraints on this transformation it is possible to maintain certain characteristics of the original graticule or to introduce new properties. Wagner derived three projection systems with a pole line from Mollweide's pseudocylindrical equal-area projection. The first projection system maintains the equal-area property of the parent projection (Wagner IV), the second allows us to specify the areal distortion on a chosen parallel (Wagner V), the third has equally spaced parallels (Wagner VI). From each of the three general projection systems it is possible to derive an unlimited number of projections through adjustment of the parameters in the transformation formulas.

For Wagner's fourth projection it is possible to adjust the ratio of the axes and the length of the pole line respectively by means of the parameters p and c. For $p = 0.5$ and $c = 0.5$ a projection is obtained with a correct ratio of the axes $(2/1)$ and a pole line half the length of the equator. This graticule was independently proposed by R. V. Putnins in 1934 and is known as Putnins P2′. Werenskiold also developed an identical graticule except that he enlarged it to achieve true scale along the equator. Therefore he multiplied the x- and y-coordinates of Wagner's projection by a factor of 1.158 62. It follows that the areal scale becomes 1.342 40 in every point of the graticule (Snyder, 1977). Thus Werenskiold's projection is no longer equal-area although the areal scale relations between different parts of the map remain unaltered.

Wagner's fourth projection does not show the excessive shearing in the higher latitudes which characterizes Mollweide's projection. However, the introduction of a pole line leads to a N–S stretching of the equatorial areas. To avoid this it is necessary to give up the equal-area property (Wagner V, Wagner VI).

Transformation formulas

$$x = R\frac{n\lambda}{\pi}\frac{2\sqrt{2}}{\sqrt{nm}}\cos\psi \qquad\qquad y = R\frac{\sqrt{2}}{\sqrt{nm}}\sin\psi$$

$$2\psi + \sin 2\psi = \pi m \sin\phi$$

where $m = \dfrac{2\arccos c + \sin(2\arccos c)}{\pi}$, $n = \dfrac{\sqrt{1-c^2}}{2p}$ and $p = \dfrac{\text{central meridian}}{\text{equator}}$, $c = \dfrac{\text{pole line}}{\text{equator}}$.

After simplification of the above general formulas and scale enlargement the following transformation formulas are obtained for Werenskiold's third projection:

$$x = R\lambda \cos\psi \qquad y = R\frac{\pi}{\sqrt{3}}\sin\psi$$

The auxiliary angle ψ is derived as for Wagner IV.

Distortion characteristics

Wagner IV	$D_{ar} = 0.00$	$D_{an} = 30.4$	$D_{ab} = 0.37$
	$D_{arc} = 0.00$	$D_{anc} = 32.1$	$D_{abc} = 0.42$
Werenskiold III	$D_{ar} = 0.34$	$D_{an} = 30.4$	$D_{ab} = 0.39$
	$D_{arc} = 0.34$	$D_{anc} = 32.1$	$D_{abc} = 0.44$

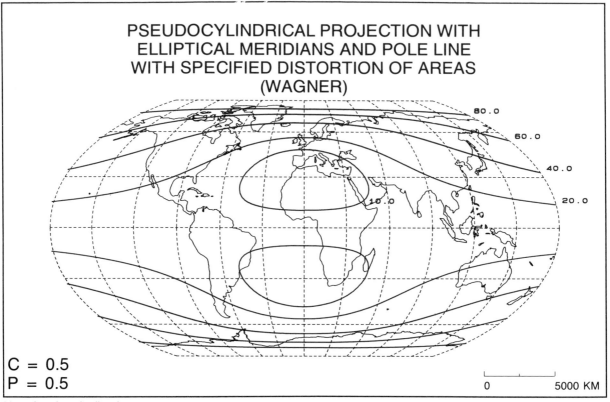

PSEUDOCYLINDRICAL PROJECTION WITH
ELLIPTICAL MERIDIANS AND POLE LINE
WITH SPECIFIED DISTORTION OF AREAS
(WAGNER)

C = 0.5
P = 0.5

0   5000 KM

Lines of equal angular distortion

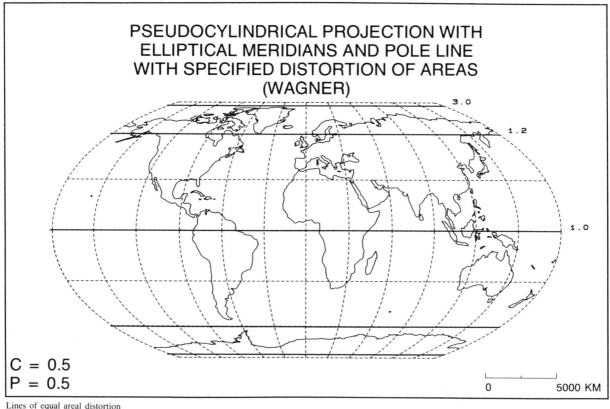

PSEUDOCYLINDRICAL PROJECTION WITH
ELLIPTICAL MERIDIANS AND POLE LINE
WITH SPECIFIED DISTORTION OF AREAS
(WAGNER)

C = 0.5
P = 0.5

0   5000 KM

Lines of equal areal distortion

6.23 Pseudocylindrical projection with elliptical meridians and pole line with specified distortion of areas (Wagner)

Other name(s):   Wagner V projection

Author(s):   Karlheinz Wagner (1962)

Description

As described in Section 2.4, a projection may be transformed by the introduction of parameters in the transformation formulas. The possibilities of this technique are unlimited, but it is especially interesting to define a transformation under certain constraints. Wagner developed a transformation which introduced a pole line without altering the areal distortion of the original projection. Applied to an equal-area projection the resulting projection remains equal-area (Wagner I, Wagner IV). A disadvantage of Wagner's pseudocylindrical equal-area projections with pole line is the N–S stretching of the equatorial areas which is combined with a compression in the same direction in the higher latitudes to assure the equal-area property. Therefore Wagner defined another transformation which allows to introduce an areal distortion that increases with increasing latitude. Moreover, this increase may be controlled by specifying an areal distortion σ_1 on a parallel ϕ_1. This transformation is explained in more detail in Section 2.4.

Wagner's fifth projection is obtained by applying the transformation to Mollweide's pseudocylindrical equal-area projection. In the resulting transformation formulas four independent parameters allow for the adjustment of the ratio of the axes (p), the length of the pole line (c) and the amount of areal distortion σ_1, ϕ_1. Wagner accepts a maximum areal distortion of 20 per cent over the area that is to be represented. For world maps he chose $p = 0.5$, $c = 0.5$, $\sigma_1 = 1.2$ and $\phi_1 = 60°$. Hence the new projection has a correct ratio of the axes (2/1), a pole line half the length of the equator and an areal distortion of 20 per cent at a latitude of 60°. This implies that almost all populated areas of the world are represented with less than 20 per cent areal distortion.

Wagner also applied the described transformation to Sanson's sinusoidal projection. Both projections have an areal distortion that is much lower than for pseudocylindrical projections with equally spaced parallels and pole line. Moreover they do not show the N–S stretching in the lower latitudes and the N–S compression in the higher latitudes which characterizes the pseudocylindrical equal-area projections with pole line.

Transformation formulas

$$x = R \frac{n\lambda}{\pi} \frac{2\sqrt{2}}{\sqrt{nm_1 m_2}} \cos \psi \qquad\qquad y = R \frac{\sqrt{2}}{\sqrt{nm_1 m_2}} \sin \psi$$

$$2\psi + \sin 2\psi = \pi m_1 \sin (m_2 \phi)$$

$$m_1 = \frac{2\arccos c + \sin(2\arccos c)}{\pi \sin(m_2(\pi/2))} \qquad m_2 = \frac{\arccos(\sigma_1 \cos \phi_1)}{\phi_1} \qquad n = \frac{\sqrt{1 - c^2}}{2p}$$

where $p = \dfrac{\text{central meridian}}{\text{equator}}$, $c = \dfrac{\text{pole line}}{\text{equator}}$ and $\sigma_1 =$ areal distortion on the parallels $\pm \phi_1$.

Note that for $m_2 = 1$ the transformation formulas of Wagner's pseudocylindrical equal-area projection with elliptical meridians and a pole line are obtained. The latter is a special case of the projection just described which maintains the equal-area property of the original projection.

Distortion characteristics

$D_{ar} = 0.11$	$D_{an} = 25.4$	$D_{ab} = 0.30$
$D_{arc} = 0.15$	$D_{anc} = 27.1$	$D_{abc} = 0.34$

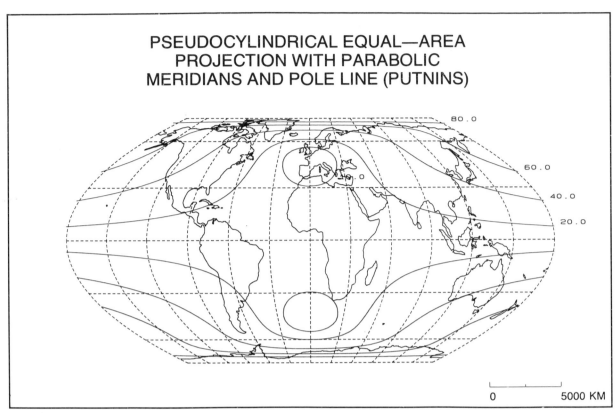

PSEUDOCYLINDRICAL EQUAL—AREA PROJECTION WITH PARABOLIC MERIDIANS AND POLE LINE (PUTNINS)

Lines of equal angular distortion

6.24 Pseudocylindrical equal-area projection with parabolic meridians and pole line (Putnins)

Other name(s): Putnins P4′ projection/Werenskiold I projection

Author(s). Reinholds V. Putnins (1934)/W. Werenskiold (1944)

Description

This projection is the equal-area version of Putnins P3′. Apart from the different scale factor, meridians, pole line and equator are identical on both projections. The main difference between both graticules is the spacing of the parallels which on Putnins P4′ decreases toward the poles to ensure the equal-area property.

The projection can be compared with Eckert's equal-area projections with pole line since it is also characterized by a correct ratio of the axes and a pole line half the length of the equator. Although the nature of the meridians is different the graticule of Putnins P4′ shows much resemblance to the graticule of Eckert VI (with sinusoidal meridians).

The meridians on the latter converge more sharply, but the difference is hardly noticeable.

In 1944 W. Werenskiold proposed independently a projection which can be considered as a modification of Putnins P4′ with true scale along the equator (Werenskiold I). This leads to an enlargement of the graticule which has no effect on the angular distortion but introduces an areal distortion which is constant throughout the whole graticule. Therefore the scale relations between different parts of the map remain unaltered. Nevertheless the projection is no longer equal-area. The areal scale is 1.309 in every point on the map so both areal distortion parameters have the value 0.309.

Transformation formulas

$$x = 2\sqrt{\frac{0.6}{\pi}} R\lambda \frac{\cos \psi}{\cos \frac{1}{3}\psi}$$

$$y = 2\sqrt{1.2\pi} R \sin \frac{\psi}{3}$$

$$\sin \psi = \frac{5\sqrt{2}}{8} \sin \phi$$

For Werenskiold's first projection the transformation formulas become

$$x = R\lambda \frac{\cos \psi}{\cos \frac{1}{3}\psi}$$

$$y = \pi\sqrt{2} R \sin \frac{\psi}{3}$$

$$\sin \psi = \frac{5\sqrt{2}}{8} \sin \phi$$

Distortion characteristics

Putnins P4′

$D_{ar} = 0.00$	$D_{an} = 31.5$	$D_{ab} = 0.38$
$D_{arc} = 0.00$	$D_{anc} = 32.6$	$D_{abc} = 0.42$

Werenskiold I

$D_{ar} = 0.31$	$D_{an} = 31.5$	$D_{ab} = 0.40$
$D_{arc} = 0.31$	$D_{anc} = 32.6$	$D_{abc} = 0.43$

FLAT—POLAR PARABOLIC AUTHALIC PROJECTION (McBRYDE AND THOMAS)

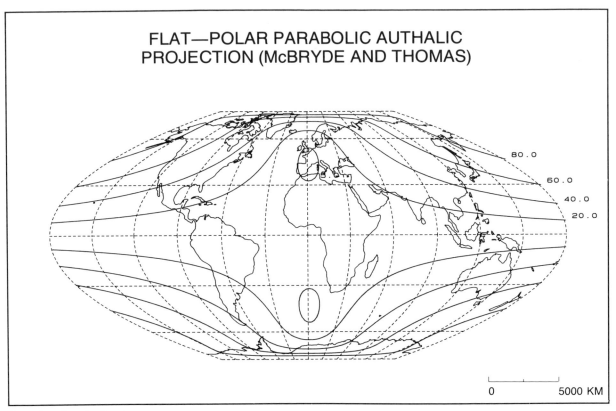

Lines of equal angular distortion

6.25 Pseudocylindrical equal-area projection with parabolic meridians and pole line (McBryde and Thomas)

Other name(s):	Flat-polar parabolic authalic projection
Author(s):	F. Webster McBryde/Paul Thomas (1949)

Description

In 1949 F. W. McBryde and P. Thomas proposed a number of pseudocylindrical equal-area projections for world statistical maps. All these projections have a pole line shorter than half the length of the equator. It was the intention of the authors to reduce the N–S stretching in the lower latitudes as well as the E–W stretching in the higher latitudes through the shortening of the pole line.

The starting point for the development of the new projections was the so-called sine series, a subgroup of the pseudocylindrical equal-area projections defined by the following general equations:

$$x = R\frac{q}{p}\lambda \, \cos\phi \, \sec\frac{\phi}{q}$$

$$y = Rp \, \sin\frac{\phi}{q}$$

The two parameters p and q allow us to construct an infinite number of map projections that satisfy these general equations. To some of these projections McBryde and Thomas applied a transformation which introduces a pole line with given length. The transformation which is comparable with Wagner's equivalent transformation principle (Section 2.4), preserves the equal-area property of the original projection.

Consider the general transformation formulas for the pseudocylindrical equal-area projection with equally divided parallels:

$$x = R\lambda \, \frac{\cos\phi}{f'(\phi)}$$

$$y = Rf(\phi)$$

After transformation this becomes

$$x = R\frac{M\lambda}{m}\left(k + \frac{\cos\alpha}{f'(\alpha)}\right)$$

$$y = RMf(\alpha)$$

where α is found by numerical solution of

$$n \, \sin\phi = kf(\alpha) + \sin\alpha$$

and with

$$m = kf'(0) + 1 \qquad n = kf\left(\frac{\pi}{2}\right) + 1 \qquad M = \sqrt{\frac{m}{n}}$$

The arbitrary parameter k allows to adjust the length of the pole line.

The flat-polar parabolic authalic projection is derived from a parabolic projection of the sine series with $p = q = 3$. The length of the pole line was chosen as one-third of the length of the equator which corresponds to $k = 0.5$.

The distortion parameters of this projection are comparable with those of Mollweide's projection and the polar areas are not represented much better than on the latter. Therefore, it can be argued whether Mollweide's projection is not to be preferred. It yields an almost identical representation of the continents with the meridians being full ellipses intersecting at the poles, while the flat-polar parabolic authalic projection shows a discontinuity at the poles as a consequence of the introduction of the pole line.

Transformation formulas

$$x = R\frac{M\lambda}{k+1}\left(k + \frac{\cos\alpha}{\cos\frac{1}{3}\alpha}\right)$$

$$y = 3RM \, \sin\frac{\alpha}{3}$$

$$n \, \sin\phi = 3k \, \sin\frac{\alpha}{3} + \sin\alpha$$

where $k = \dfrac{1}{2}$, $n = \dfrac{3}{2}k + 1 = 1.75$ and $M = \sqrt{\dfrac{k+1}{n}} = 0.925\,8$.

The value of the auxiliary angle α is calculated by numerical approximation.

Distortion characteristics

$D_{ar} = 0.00$	$D_{an} = 33.7$	$D_{ab} = 0.41$
$D_{arc} = 0.00$	$D_{anc} = 33.9$	$D_{abc} = 0.42$

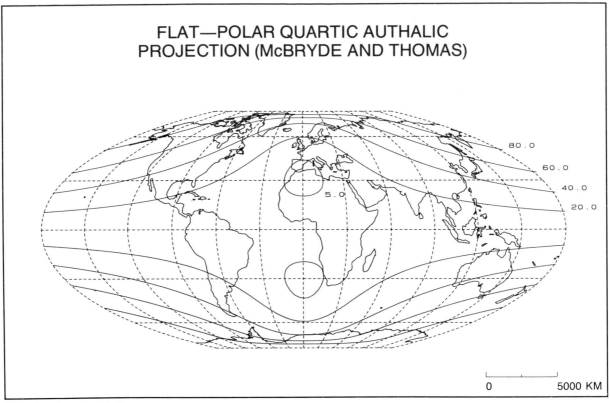

FLAT—POLAR QUARTIC AUTHALIC
PROJECTION (McBRYDE AND THOMAS)

0   5000 KM

Lines of equal angular distortion

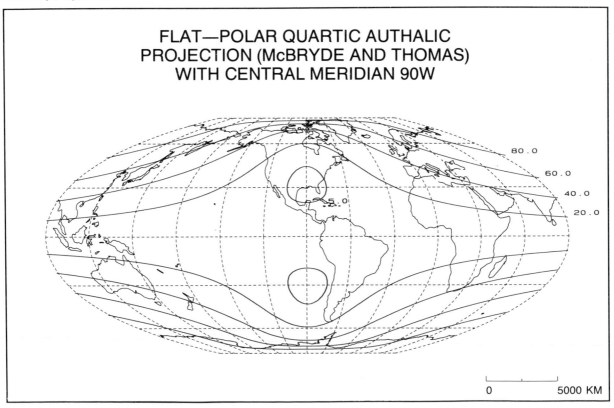

FLAT—POLAR QUARTIC AUTHALIC
PROJECTION (McBRYDE AND THOMAS)
WITH CENTRAL MERIDIAN 90W

0   5000 KM

Lines of equal angular distortion

6.26 Pseudocylindrical equal-area projection with quartic meridians and pole line (McBryde and Thomas)

Other name(s):	Flat-polar quartic authalic projection
Author(s):	F. Webster McBryde and Paul Thomas (1949)

Description

This projection with quartic meridians is, just as those with sinusoidal and parabolic meridians of McBryde and Thomas, characterized by a pole line with length one-third of the equator. It results from the application of an equivalent transformation to Adams's pseudocylindrical equal-area projection with quartic meridians. The nature of the applied transformation is explained in the description of the pseudocylindrical equal-area projection with parabolic meridians and pole line (flat-polar parabolic authalic projection).

The meridians in the transformed projection are curves of fourth degree which leads to a better representation of the polar regions than in the flat-polar projections with sinusoidal and parabolic meridians. This is reflected in a lower mean angular distortion.

Moreover the considerable elongation of the N–S distances in the lower latitudes, which is typical for pseudocylindrical equal-area projections with a pole line half the length of the equator (e.g. Eckert's equal-area projections), is remedied by the shortening of the pole line.

Transformation formulas

$$x = R \frac{M\lambda}{k+1} \left(k + \frac{\cos \alpha}{\cos \frac{1}{2}\alpha} \right) \qquad\qquad y = 2RM \sin \frac{\alpha}{2}$$

$$n \sin \phi = 2k \sin \frac{\alpha}{2} + \sin \alpha$$

where $k = \dfrac{1}{2}$, $n = k\sqrt{2} + 1 = 1.7071$ and $M = \sqrt{\dfrac{k+1}{n}} = 0.9374$.

The value of α is calculated by numerical approximation.

Distortion characteristics

$D_{ar} = 0.00$	$D_{an} = 32.1$	$D_{ab} = 0.39$
$D_{arc} = 0.00$	$D_{anc} = 33.2$	$D_{abc} = 0.42$

Recentered flat-polar quartic authalic projection with central meridian at 90° W

In order to make a comparison possible between the different projection systems in the Directory, most of them have been shown with the Greenwich meridian central. This is, however, an arbitrary choice. An American view for instance might be obtained by choosing the central meridian at 90° W.

FLAT—POLAR SINUSOIDAL AUTHALIC PROJECTION (McBRYDE AND THOMAS)

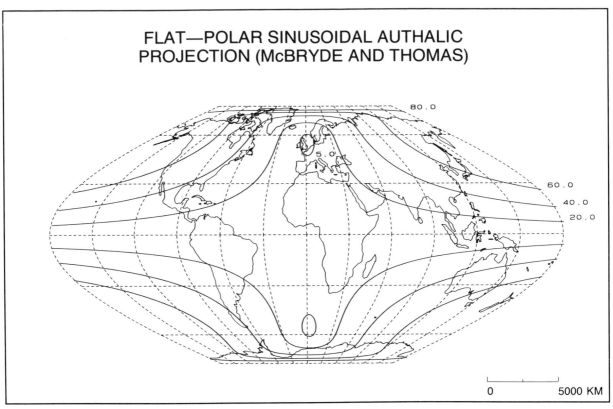

Lines of equal angular distortion

6.27 Pseudocylindrical equal-area projection with sinusoidal meridians and pole line (McBryde and Thomas)

Other name(s): Flat-polar sinusoidal authalic projection

Author(s): F. Webster McBryde and Paul Thomas (1949)

Description

This projection forms part of a series of projections that were developed by McBryde and Thomas and are characterized by a pole line shorter than half the length of the equator. The main idea behind the development of these projections was the reduction of the N–S stretching in the lower latitudes which is typical for equal-area projections with a pole line. This may be achieved in two ways:
1. By giving up the equal-area property;
2. By a shortening of the pole line.
McBryde and Thomas chose the second solution. As is explained in the description of their pseudocylindrical equal-area projection with parabolic meridians and pole line (flat-polar parabolic authalic projection), they applied an equivalent transformation to the so-called sine series of pseudocylindrical equal-area projections. This resulted in an equal-area projection system with a pole line.

This projection was derived from Sanson's sinusoidal projection and has a pole line one third the length of the equator. The mean angular distortion of this projection is higher than for the flat-polar projections with parabolic and quartic meridians. This illustrates that sinusoidal meridians are less suited for world map projections with small E–W extension at higher latitudes.

Transformation formulas

$$x = R\frac{M\lambda}{k+1}(k + \cos\alpha) \qquad\qquad y = RM\alpha$$

$$n\sin\phi = k\alpha + \sin\alpha$$

where $k = \dfrac{1}{2}$, $n = \dfrac{1}{2}(k\pi + 2) = 1.7854$ and $M = \sqrt{\dfrac{k+1}{n}} = 0.9166$.

The value of α is calculated by numerical approximation.

Distortion characteristics

$D_{ar} = 0.00$	$D_{an} = 35.1$	$D_{ab} = 0.42$
$D_{arc} = 0.00$	$D_{anc} = 34.6$	$D_{abc} = 0.42$

ECKERT VI PROJECTION

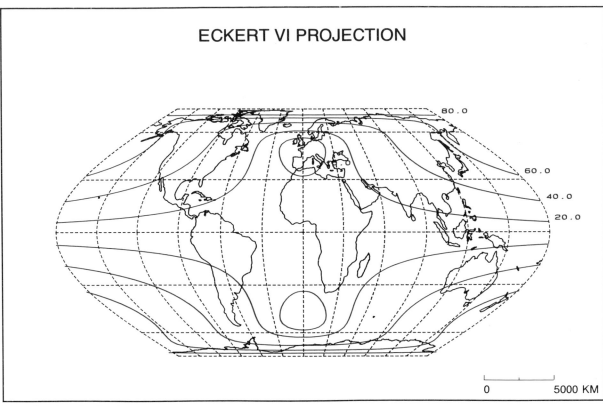

0   5000 KM

Lines of equal angular distortion

6.28 Pseudocylindrical equal-area projection with sinusoidal meridians and pole line (Eckert)

Other name(s): Eckert VI projection

Author(s): Max Eckert (1906)

Description

The projection is an equal-area version of Eckert's pseudocylindrical projection with equally spaced parallels, sinusoidal meridians and pole line (Eckert V). The realization of the equal-area property implies a decreasing parallel spacing with increasing latitude. This leads to an elongation of the N–S distances in the lower latitudes typical for a pseudocylindrical equal-area projection with a correct ratio of the axes and a pole line half the length of the equator. It was this consideration that brought McBryde and Thomas (1949) to develop their so-called 'flat-polar projections', a series of equal-area projections with a pole line shorter than half the length of the equator.

In spite of the above mentioned shortcoming, Eckert's sixth projection has a more balanced distortion pattern than Eckert's fourth. The graticule of the projection is almost identical to that of Kavraisky's sixth projection, which in its turn is a special case of Wagner's pseudocylindrical equal-area projection with sinusoidal meridians and pole line.

Transformation formulas

$$x = \frac{r\lambda}{\pi}(1 + \cos \psi)$$

$$y = \frac{2r}{\pi}\psi$$

$$\psi + \sin \psi = \frac{\pi + 2}{2} \sin \phi$$

$$r = R\frac{\pi}{\sqrt{\pi + 2}}$$

where r = the length of the semi minor axis of the planisphere.

The transcendental equation that expresses the relation between the geographical latitude ϕ and the auxiliary angle ψ is solved numerically.

Distortion characteristics

$D_{ar} = 0.00$	$D_{an} = 32.4$	$D_{ab} = 0.39$
$D_{arc} = 0.00$	$D_{anc} = 32.9$	$D_{abc} = 0.42$

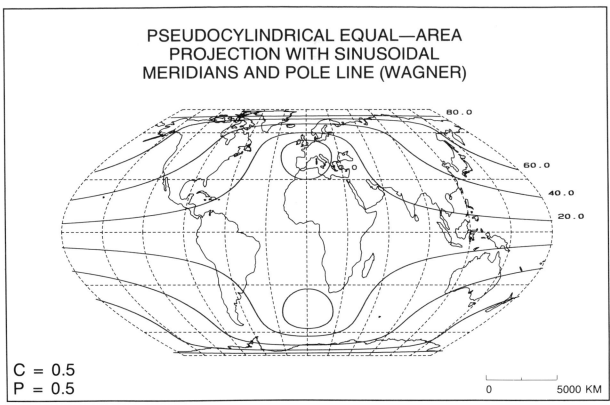

PSEUDOCYLINDRICAL EQUAL—AREA
PROJECTION WITH SINUSOIDAL
MERIDIANS AND POLE LINE (WAGNER)

C = 0.5
P = 0.5

0   5000 KM

Lines of equal angular distortion

6.29 Pseudocylindrical equal-area projection with sinusoidal meridians and pole line (Wagner)

Other name(s):   Wagner I projection/Kavraisky VI projection/
Werenskiold II projection

Author(s).   Karlheinz Wagner (1962)/V.V. Kavraisky (1936)/
W. Werenskiold (1944)

Description

This projection system results from the application of Wagner's equivalent transformation principle (Section 2.4) to Sanson's sinusoidal projection. By means of the two parameters p and c in the general transformation formulas of this system it is possible to adjust the ratio of the axes and the length of the pole line respectively. Wagner chose $p = 0.5$ and $c = 0.5$ and obtained a graticule almost identical to Eckert's pseudocylindrical equal-area projection with sinusoidal meridians and pole line (Eckert VI). Both projections have very similar distortion patterns. The only important difference between them is the way in which coordinates are calculated. For Eckert's sixth projection the relation between the auxiliary angle ψ and the geographical latitude ϕ is given by a transcendental equation while for Wagner's projection the value of ψ, that corresponds to a given value of ϕ, can be calculated directly.

Wagner developed the described projection system in 1932 together with two other pseudocylindrical projections with sinusoidal meridians and a pole line which were also obtained through transformation of Sanson's sinusoidal projection. The special case of this projection with $p = c = 0.5$ was developed independently by Kavraisky in 1936 and is known as Kavraisky VI. Also Werenskiold's second projection (Werenskiold, 1944) is identical although he introduced a scale factor to maintain the principal scale along the equator. This scale enlargement has no effect on the angular distortion but results in an areal distortion which is constant throughout the whole graticule.

Transformation formulas

$$x = R \frac{n\lambda}{\sqrt{nm}} \cos \psi \qquad\qquad y = R \frac{\psi}{\sqrt{nm}}$$

$$\sin \psi = m \sin \phi$$

where $m = \sqrt{1 - c^2}$, $n = \dfrac{\arcsin m}{p\pi}$

For Werenskiold's second projection the transformation formulas become

$$X = R\lambda \cos \psi \qquad\qquad y = 3R \frac{\psi}{2}$$

$$\sin \psi = \frac{\sqrt{3}}{2} \sin \phi$$

and where $p = \dfrac{\text{central meridian}}{\text{equator}}$, $c = \dfrac{\text{pole line}}{\text{equator}}$.

Distortion characteristics

Wagner I	$D_{ar} = 0.00$	$D_{an} = 31.9$	$D_{ab} = 0.38$
	$D_{arc} = 0.00$	$D_{anc} = 32.7$	$D_{abc} = 0.42$
Werenskiold II	$D_{ar} = 0.30$	$D_{an} = 31.9$	$D_{ab} = 0.40$
	$D_{arc} = 0.30$	$D_{anc} = 32.7$	$D_{abc} = 0.43$

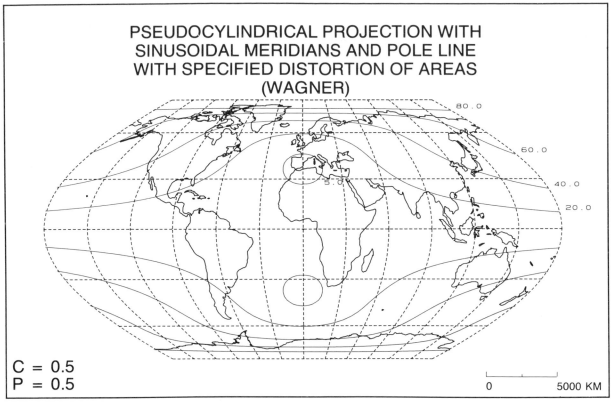

PSEUDOCYLINDRICAL PROJECTION WITH
SINUSOIDAL MERIDIANS AND POLE LINE
WITH SPECIFIED DISTORTION OF AREAS
(WAGNER)

C = 0.5
P = 0.5

0   5000 KM

Lines of equal angular distortion

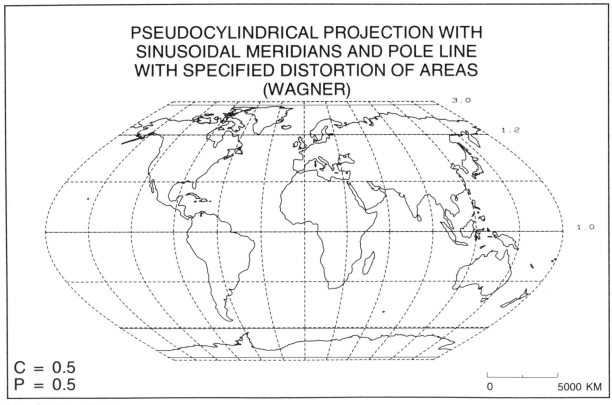

PSEUDOCYLINDRICAL PROJECTION WITH
SINUSOIDAL MERIDIANS AND POLE LINE
WITH SPECIFIED DISTORTION OF AREAS
(WAGNER)

C = 0.5
P = 0.5

0   5000 KM

Lines of equal areal distortion

6.30 Pseudocylindrical projection with sinusoidal meridians and pole line with specified distortion of areas (Wagner)

Other name(s):	Wagner II projection
Author(s):	Karlheinz Wagner (1962)

Description

Every pseudocylindrical equal-area projection with a pole line elongates the N–S distances in the lower latitudes while the polar areas are compressed in this direction. This may be avoided by giving up the equal-area property.

Wagner defined a transformation which allows to introduce an areal distortion that gradually increases from $\sigma = 1.0$ on the equator to $\sigma = \infty$ at the poles. Moreover, it is possible to control this latitudinal increase by specifying an areal distortion σ_1 on a parallel ϕ_1. Wagner proposed an areal distortion $\sigma = 1.2$ for $\phi_1 = 60°$. This allows to represent the largest part of the *oikumene* (inhabited world) with less than 20 per cent areal distortion. He first applied this transformation to Sanson's sinusoidal projection and chose a

correct ratio of the axes ($p = 0.5$) and a pole line half the length of the equator ($c = 0.5$). The result is a projection with a low areal distortion (much lower than for Wagner's pseudocylindrical projections with equally spaced parallels) and less scale variation along the central meridian than on the equal-area version (Wagner's pseudocylindrical equal-area projection with sinusoidal meridians and pole line).

Wagner also applied the described transformation to Mollweide's projection. This led to a projection with similar distortion characteristics (Wagner's pseudocylindrical projection with elliptical meridians and pole line with specified distortion of areas).

Transformation formulas

$$x = R\frac{n\lambda}{\sqrt{nm_1m_2}}\cos\psi \qquad\qquad y = R\frac{\psi}{\sqrt{nm_1m_2}}$$

$$\sin\psi = m_1\sin(m_2\phi)$$

where $m_1 = \dfrac{\sqrt{1-c^2}}{\sin(m_2(\pi/2))}$, $m_2 = \dfrac{\arccos(\sigma_1\cos\phi_1)}{\phi_1}$ and $n = \dfrac{\arcsin\left(\sqrt{1-c^2}\right)}{p\pi}$

and where $p = \dfrac{\text{central meridian}}{\text{equator}}$, $c = \dfrac{\text{pole line}}{\text{equator}}$, $\sigma_1 = $ areal distortion on the parallels $\pm\ \phi_1$.

Distortion characteristics

$D_{ar} = 0.11$	$D_{an} = 26.9$	$D_{ab} = 0.31$
$D_{arc} = 0.15$	$D_{anc} = 27.7$	$D_{abc} = 0.34$

NELL—HAMMER PROJECTION

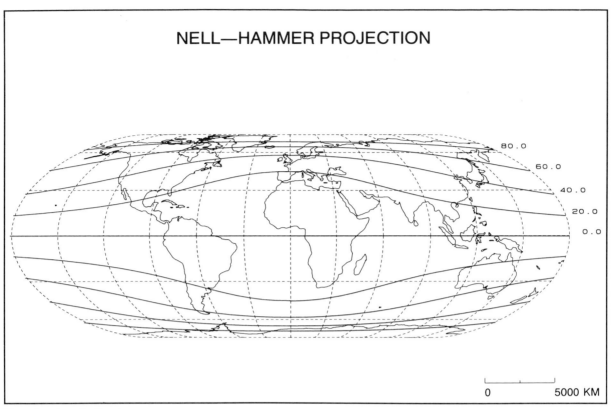

80.0

60.0

40.0

20.0

0.0

0   5000 KM

Lines of equal angular distortion

6.31 Pseudocylindrical equal-area projection with pole line (Hammer)

Other name(s): Nell–Hammer projection

Author(s): E. Hammer (1900)/A. M. Nell (1890)

Description

A very important group among the pseudocylindrical projections with pole line are the projections which are defined as the arithmetic mean of a cylindrical projection and a pointed-polar pseudocylindrical projection. In most cases the cylindrical equidistant projection is used as one of the two parent projections, e.g. Eckert's projections, Winkel's projections, Putnins' projections. Nevertheless, a few authors also used the cylindrical equal-area projection.

When the resulting projection has to be equal-area there are two possibilities. Either the x-coordinates or the y-coordinates of the two parent projections may be averaged. Then the other coordinate of the pseudocylindrical equal-area projection with equally divided parallels is derived from the equal-area condition (Section 1.2, equation 33):

$$x = R^2 \lambda \frac{\cos \phi}{dy/d\phi}$$

$$y = Rf(\phi)$$

Hammer formed a pseudocylindrical equal-area projection by averaging the x-coordinate of the cylindrical equal-area projection and Sanson's sinusoidal projection. Hence

$$x = \frac{R\lambda}{2}(1 + \cos \phi)$$

Then the y-coordinate follows from

$$y = R^2 \int_0^\phi \frac{\lambda \cos \phi}{x} d\phi$$

The original idea for the development of this projection came from A. M. Nell. Therefore it is known as the Nell–Hammer projection. The projection has no real practical value. The unusual ratio of the axes (the equator is 2.75 times the length of the central meridian) is not pleasing, the E–W stretching in the higher latitudes is extreme. Nevertheless the projection illustrates very well the principle of averaging two projections, thereby maintaining the equal-area property. It is one of the oldest known projections to accomplish this.

The method of constructing an equal-area projection by averaging the coordinates of two existing projections may be extended by giving a different weight to the coordinates of each parent projection. Foucaut used this approach in averaging the y-coordinates of the cylindrical equal-area and the sinusoidal projection. Tobler presented geometrical means of x- or y-coordinates of the same two projections (Snyder, 1977).

Transformation formulas

$$x = \frac{R\lambda}{2}(1 + \cos \phi)$$

$$y = 2R\left(\phi - \tan \frac{\phi}{2}\right)$$

Distortion characteristics

$D_{ar} = 0.00$	$D_{an} = 30.9$	$D_{ab} = 0.42$
$D_{arc} = 0.00$	$D_{anc} = 33.6$	$D_{abc} = 0.50$

ECKERT II PROJECTION

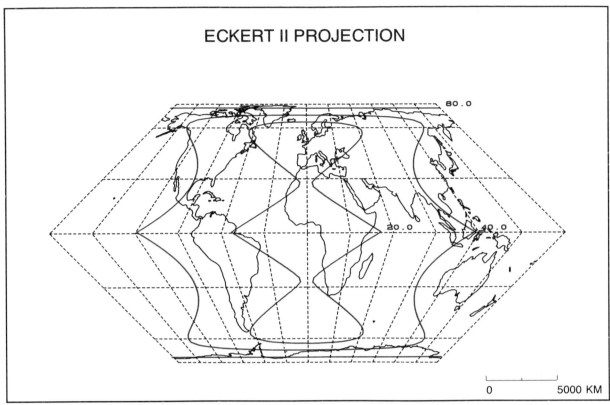

Lines of equal angular distortion

6.32 Pseudocylindrical equal-area projection with rectilinear meridians and pole line (Eckert)

Other name(s): Eckert II projection

Author(s): Max Eckert (1906)

Description

Eckert's second projection is an equal-area version of Eckert I. Since Eckert's first projection is total area true, the equal-area property is accomplished by decreasing the spacing of the parallels so that the area bounded by two arbitrary parallels equals the area of the corresponding zone on the globe. Apart from this difference both graticules are completely identical.

The graticule of Eckert's second projection is very easy to construct. Nevertheless, this argument does not justify its use since there is a discontinuity of meridians at the equator which yields a considerable distortion of the equatorial areas. Moreover, as a result of the equal-area property, the N–S distances are elongated in the lower latitudes while the polar areas are compressed in the same direction. Compared with Eckert's two other equal-area projections (Eckert IV, Eckert VI) Eckert II has a different distortion pattern as a consequence of the discontinuity at the equator. This is reflected in the distortion parameters which have relatively high values for Eckert's second projection.

Transformation formulas

$$x = \frac{\lambda}{\pi}(2r - |y|) \qquad y = 2r - \sqrt{4r^2 - 2\pi \sin |\phi|} \quad \phi \geq 0$$

$$r = R\sqrt{\frac{2}{3}\pi}$$

where r = half the length of the minor axis.

The expression for the y-coordinate applies only to the Northern Hemisphere. The Southern Hemisphere is obtained by mirroring about the equator.

Distortion characteristics

$D_{ar} = 0.00$	$D_{an} = 38.2$	$D_{ab} = 0.46$
$D_{arc} = 0.00$	$D_{anc} = 35.3$	$D_{abc} = 0.45$

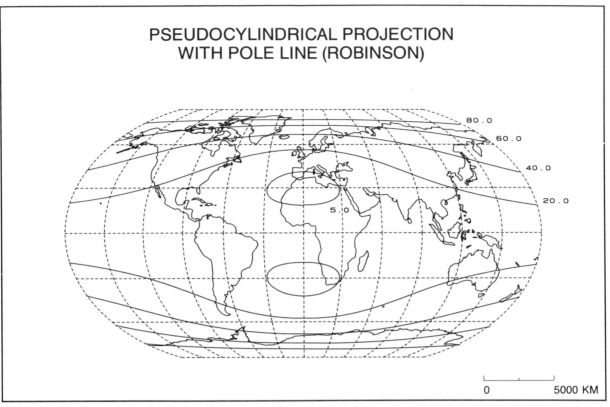

PSEUDOCYLINDRICAL PROJECTION
WITH POLE LINE (ROBINSON)

Lines of equal angular distortion

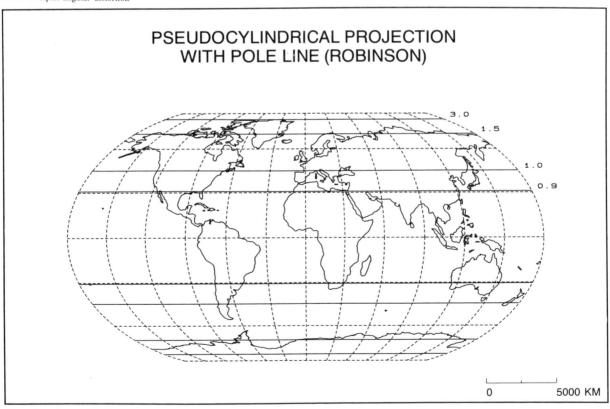

PSEUDOCYLINDRICAL PROJECTION
WITH POLE LINE (ROBINSON)

Lines of equal areal distortion

6.33 Pseudocylindrical projection with pole line (Robinson)

Other name(s):	Robinson projection
Author(s):	Arthur H. Robinson (1974)

Description

In 1974 A. H. Robinson described a new map projection especially designed for general-purpose world maps. The motivation behind the development of this projection was that none of the existing projections for world maps satisfied the requirements which were pointed out by the ordering company (Rand McNally). The set of given constraints led to a non-equal-area, pseudocylindrical projection, with a pole line, and with parallels evenly divided by the meridians. Moreover, it was decided that the ratio between the length of the equator and the length of the central meridian would not exceed 2 : 1, and that the length of the pole lines would be more than half the length of the equator. Within this framework the new projection was developed by an iterative plotting process that was repeated until the shapes of the land masses, with the exception of the higher latitudes, were as realistic as they

could be. Therefore the final map is to a large extent the result of the designer's experience and was not mathematically derived. Knowing that the length of the equator is 0.8487 times the circumference of a sphere of equal area, and that the length of the central meridian is 0.5072 times the length of the equator, Table 6.3 unambiguously defines the graticule of the new projection. The map somehow resembles Kavraisky's pseudocylindrical projection with equally spaced parallels, elliptical meridians and pole line (Kavraisky VII). However, through a gradual decrease of the parallel spacing in the higher latitudes, the areal distortion is considerably less on Robinson's map, while the angular distortion is only slightly higher than on Kavraisky's graticule. Hence Robinson's map is more pleasing, especially because the land masses near the poles are less exaggerated.

Table 6.3   Spacings of Robinson's graticule (after A. H. Robinson, 1974)

ϕ (°)	Distance of parallels from equator ($0° - 90° = 1.0$)	Length of parallels (equator = 1.0)	ϕ (°)	Distance of parallels from equator ($0° - 90° = 1.0$)	Length of parallels (equator = 1.0)
90°	1.0000	0.5322	40°	0.4958	0.9216
85°	0.9761	0.5722	35°	0.4340	0.9427
80°	0.9394	0.6213	30°	0.3720	0.9600
75°	0.8936	0.6732	25°	0.3100	0.9730
70°	0.8435	0.7186	20°	0.2480	0.9822
65°	0.7903	0.7597	15°	0.1860	0.9900
60°	0.7346	0.7986	10°	0.1240	0.9954
55°	0.6769	0.8350	5°	0.0620	0.9986
50°	0.6176	0.8679	0°	0.0000	1.0000
45°	0.5571	0.8962			

Transformation formulas

As stated, no mathematical formulas are available for Robinson's projection. However, on the basis of Table 6.3, formulas were developed by polynomial approximation to

make a distortion analysis possible.

In general, a pseudocylindrical projection with equally divided parallels and pole line can be described by

$$x = R\lambda(A_0 + A_2\phi^2 + A_4\phi^4 + \dots)$$

$$y = R(A_1\phi + A_3\phi^3 + A_5\phi^5 + \dots)$$

For a fifth-order polynomial approximation the least squares procedure gives the following coefficients:

$A_0 = 0.8507$	$A_2 = -0.1450$	$A_4 = -0.0104$
$A_1 = 0.9642$	$A_3 = -0.0013$	$A_5 = -0.0129$

Each point of the uniformly spaced graticule was weighted by the area of the grid cell surrounding the point. The fifth-order approximation gives a root mean square error of less

than 0.1 mm on a scale of 1 : 5 000 000, which means that it is accurate enough to support the development of large scale wall maps based on this graticule.

Distortion characteristics	$D_{ar} = 0.21$	$D_{an} = 21.4$	$D_{ab} = 0.26$
	$D_{arc} = 0.25$	$D_{anc} = 23.2$	$D_{abc} = 0.30$

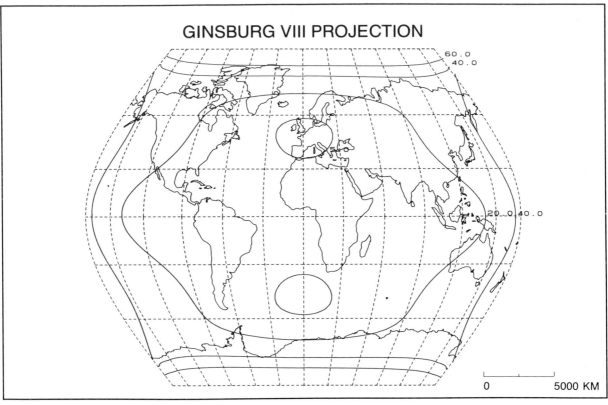

Lines of equal angular distortion

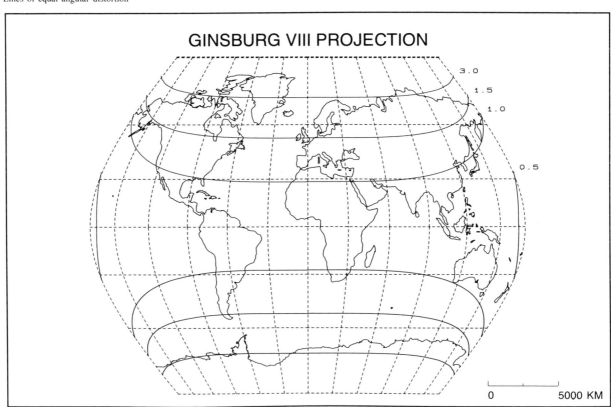

Lines of equal areal distortion

6.34 Pseudocylindrical projection (Ginsburg)

Other name(s): Ginsburg VIII projection/
TsNIGAiK pseudocylindrical projection

Author(s): G. A. Ginsburg (1949)

Description

In the Western world the interest in projection systems which are neither conformal nor equal-area is a relatively recent phenomenon. Russian cartographers like G. A. Ginsburg and T. D. Salmanova emphasized the merits of these projections 40 years ago. Ginsburg developed several polyconic and one pseudocylindrical projection using Urmaev's method of analysis (Maling, 1960). Normally the development of a new projection starts from a set of imposed geometrical constraints and some special properties and then leads to a graticule with a specific distortion pattern. Urmaev's method, on the other hand, starts with the specification of certain distortion values for particular parts of the map and yields a projection which satisfies these imposed constraints. In the description of Urmaev's third projection (Urmaev III) the method is described briefly for the simple case of a cylindrical projection.

Ginsburg's eighth projection was originally developed with a central meridian at 70° E. Urmaev's method was applied to represent the USSR in a better way than on other pseudocylindrical projections (Maling, 1960). The projection has a low angular distortion. It increases rapidly near the edges of the map only. The areal distortion, on the other hand, is relatively high.

A disadvantage of the projection is the large-scale variation along the equator. This variation leads to an unnatural ratio of the axes.

Transformation formulas

$$x = R(1 - b\phi^2)(c - d\lambda^4)\lambda \qquad y = R(\phi + a\phi^3)$$

where $a = \dfrac{1}{12}$, $b = 0.162\,388$, $c = 0.87$ and $d = 0.000\,952\,426$.

Distortion characteristics

$D_{ar} = 0.49$	$D_{an} = 20.3$	$D_{ab} = 0.29$
$D_{arc} = 0.59$	$D_{anc} = 17.5$	$D_{abc} = 0.29$

PSEUDOCONICAL PROJECTIONS

	Pole = point
Equally spaced parallels	7.1 Bonne/ Werner (cordiform projection)

From Ptolemaeus' Geographia (Francesco Berlinghieri, Florence, 1482).

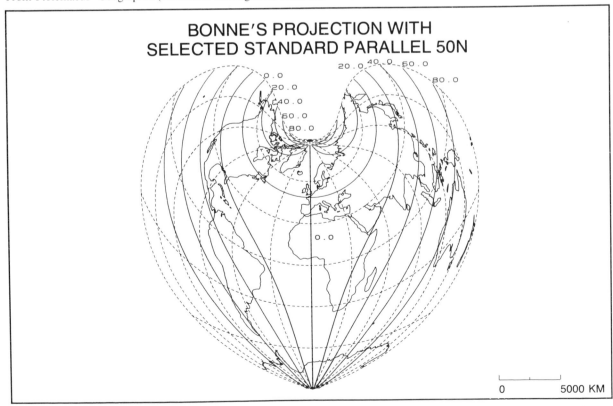

Lines of equal angular distortion

7.1 Pseudoconical equal-area projection (Bonne)

Other name(s):   Bonne's projection

Author(s):   Rigobert Bonne

Description

On a pseudoconical projection the meridians are represented by concurrent curves, the parallels by concentric arcs of circles. The centre of the pseudoconical equal-area projection is defined as the intersection of a straight central meridian with a central parallel. This parallel is constructed in the same way as the standard parallel on a conical projection and represented as a circle with radius

$$r_0 = R \tan \delta_0 \qquad \delta_0 = 90° - \phi_0$$

ϕ_0 = latitude of the central parallel

Starting from the centre of the projection the parallels are correctly spaced along the central meridian and constructed as arcs of circles of true length with the same centre as the central parallel. Each parallel is equally divided by the curved meridians.

Although the pseudoconical equal-area projection is named for R. Bonne (1727–1795), the concept has to be attributed to Ptolemaeus (second century). Ptolemaeus developed a projection with circular parallels and curved meridians for a map of the ancient world which is generally referred to as Ptolemaeus II. The difference with Bonne's projection is that Ptolemaeus represented only three parallels by arcs of true length (Thule, Syene, Anti-Meroe). These parallels were equally divided by circular meridians. Some authors stated that the equator was also involved. In that case the meridians are curves of higher degree. Although Ptolemaeus' second projection is not equal- area it approximates the equal-area projection which is obtained by representing all parallels in correct length. Therefore, it may be stated that the concept of Bonne's projection was laid down by Ptolemaeus in the second century. However, Bonne was the first to describe the pseudoconical equal-area projection mathematically in the second half of the eighteenth century. In the nineteenth century the graticule was used for topographical purposes, e.g. in France and Belgium. Today conformal projections are preferred for topographical maps, and the use of Bonne's projection became restricted to small-scale maps of continental size.

7.1.1   Pseudoconical equal-area projection ($\phi_0 = 50°$ N)

Description

The appearance of the pseudoconical equal-area projection depends on the choice of the central parallel. Minimization of the scale errors over continental area (D_{abc}) leads to a central parallel at 50° N (Fig. 7.1). Although this projection has a low mean angular distortion significant parts of the Southern Hemisphere continents (Australia, South America, Antarctica) are severely distorted. This is reflected in the high value for the global distortion parameters. Minimization of the overall scale errors (D_{ab}) yields the equator as central parallel. In that case all parallels become straight lines and Sanson's sinusoidal projection is obtained (Sanson's pseudocylindrical equal-area projection with sinusoidal meridians). The Southern Hemisphere has now become a mirror image to the Northern Hemisphere (the level of symmetry has been raised) which explains that Sanson's graticule has the lowest overall scale error (Fig. 7.1).

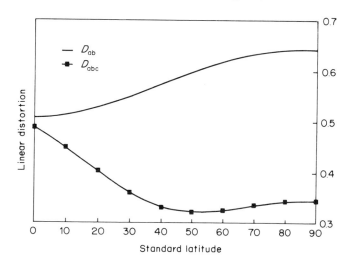

Fig. 7.1   Linear distortion as a function of standard latitude for Bonne's pseudoconical equal-area projection.

Apianus' cordiform World Map (Ingelstadt, 1530).

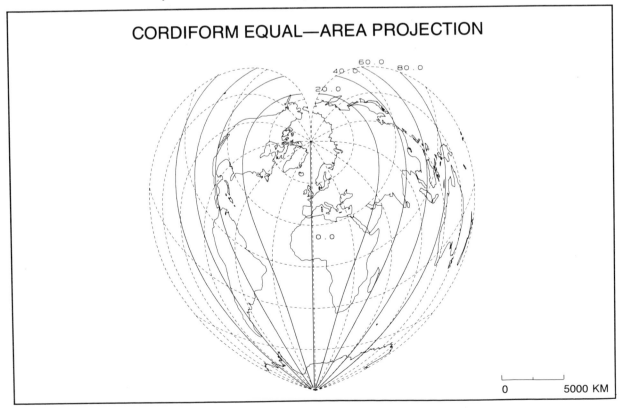

Lines of equal angular distortion

Transformation formulas

$$x = r \sin \theta$$

$$y = R \cot \phi_0 - r \cos \theta$$

$$\theta = \frac{R\lambda \cos \phi}{r}$$

$$r = R[\cot \phi_0 - (\phi - \phi_0)]$$

$$\phi_0 = \text{latitude of the central parallel}$$

Distortion characteristics

$D_{ar} = 0.00$	$D_{an} = 43.2$	$D_{ab} = 0.60$
$D_{arc} = 0.00$	$D_{anc} = 25.8$	$D_{abc} = 0.32$

7.1.2 *Pseudoconical equal-area projection* ($\phi_0 = 90°$ *N*)

Other name(s):   Cordiform equal-area projection/
Werner's second projection/
Werner's equivalent projection

Author(s):   Johannes Werner (1514)

Description

When the pole is taken as the centre of the projection the graticule becomes typically heart-shaped (the pole is now the centre of the parallels) and the projection is therefore called 'cordiform'. The asymmetry between the two hemispheres is extreme and the global distortion parameters reach their highest value. It is supposed that the first cordiform map was constructed by Oronce Fine (1519), while the concept is attributed to Johannes Werner (1514). Fine's map however was not published until 1534. The oldest published map of this type is the map made by Petrus Apianus (1530). It is based on Fine's map although it is not known whether its use was authorized by Fine or not (Shirley, 1984). The cordiform projections were very popular in the sixteenth century but were seldom used afterwards.

Distortion characteristics

$D_{ar} = 0.00$	$D_{an} = 45.7$	$D_{ab} = 0.64$
$D_{arc} = 0.00$	$D_{anc} = 27.8$	$D_{abc} = 0.34$

8

CYLINDRICAL PROJECTIONS

	Pole = line	
Equally spaced parallels	*8.1 Cylindrical equidistant projection* plate *carrée* equi-rectangular projection	(one standard parallel) (two standard parallels)
Decreasing parallel spacing	*8.2 Cylindrical equal-area projection* Lambert Behrmann Peters 8.3 Pavlov	(one standard parallel) (two standard parallels at 30°) (two standard parallels at 45°)
Increasing parallel spacing	8.4 Mercator 8.5 Miller I/Miller II *8.6 Cylindrical stereographic projection* Braun BSAM Gall 8.7 Urmaev III	 (one standard parallel) (two standard parallels at 30°) (two standard parallels at 45°)

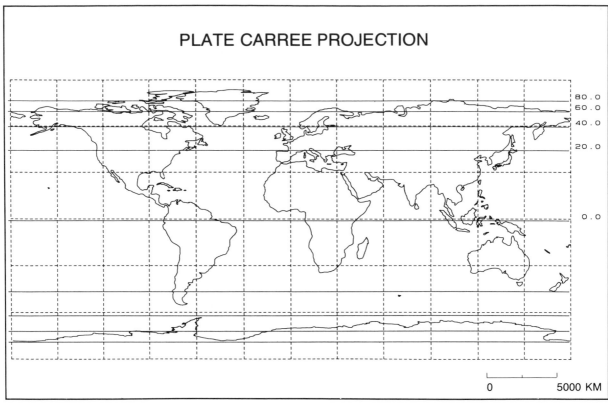

Lines of equal angular distortion

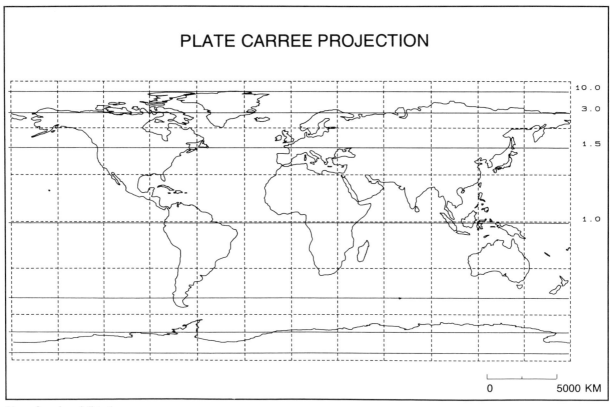

Lines of equal areal distortion

8.1 Cylindrical equidistant projection

8.1.1 Cylindrical equidistant projection with one standard parallel

Other name(s): Plate Carrée/Simple cylindrical projection

Author(s): Eratosthenes (third century BC)

Description

The plate *carrée* is the most simple cylindrical projection. The equator and all the meridians are represented as equidistant straight lines. This results in a square graticule which explains its name. Since the meridians do not converge, the E–W stretching increases rapidly toward the poles. At a latitude of 60° the parallels are already doubled in length, and since the distance between the parallels remains constant this leads to a considerable areal and angular distortion at higher latitudes. The projection is therefore not suited for the development of world maps. However, the graticule is very easy to construct, and in earlier days this certainly contributed to the propagation of its use.

The cylindrical equidistant projection with one standard parallel (the equator) is one of the oldest projection systems. It is attributed to Eratosthenes (third century BC). The transverse aspect (Cassini's projection) was widely used for topographical mapping from the mid-eighteenth century until the beginning of the twentieth century. Today conformal projections are preferred for this purpose.

The projection is included in this directory because of its historical importance and because it underlies the construction of many projections with a pole line (e.g. Eckert's projections).

Transformation formulas

$$x = R\lambda$$

$$y = R\phi$$

Distortion characteristics

$D_{ar} = 0.55$	$D_{an} = 16.8$	$D_{ab} = 0.27$
$D_{arc} = 0.77$	$D_{anc} = 21.0$	$D_{abc} = 0.38$

MILLER'S EQUIRECTANGULAR PROJECTION

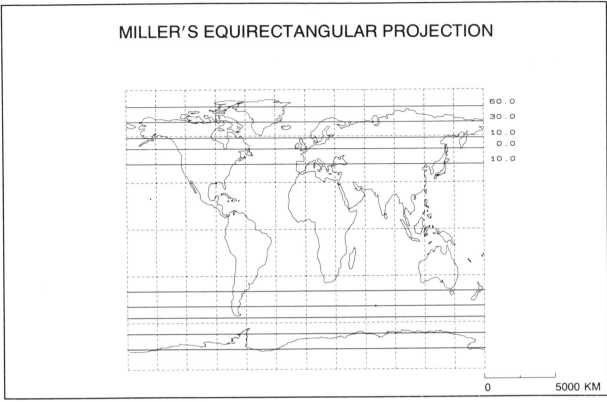

60.0
30.0
10.0
0.0
10.0

0   5000 KM

Lines of equal angular distortion

MILLER'S EQUIRECTANGULAR PROJECTION

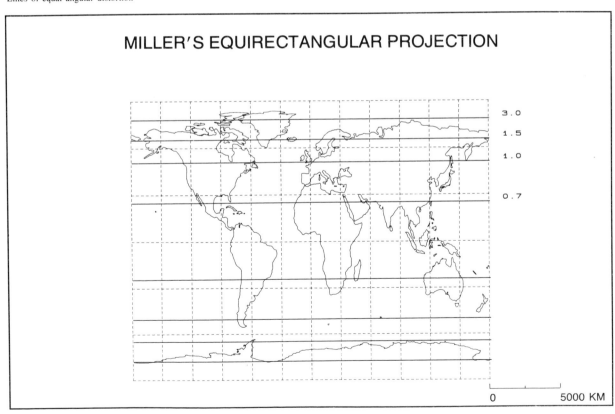

3.0
1.5
1.0
0.7

0   5000 KM

Lines of equal areal distortion

8.1   Cylindrical equidistant projection (CONTINUED)

8.1.2   *Cylindrical equidistant projection with two standard parallels*

Other name(s):   Equi-rectangular projection

Author(s).   Marinus of Tyre (± 100 BC)

Description

If instead of the equator an arbitrary parallel is represented in correct length the square graticule of the cylindrical equidistant projection with one standard parallel (plate *carrée*), is transformed into a rectangular graticule. The shape of the grid depends on the choice of the two standard parallels which lie symmetric about the equator. The plate *carrée* can be considered as a limiting case of this general projection system where the two standard parallels coincide with the equator.

R. Miller (1949) developed a cylindrical equidistant projection with standard latitude $\phi_0 = 50° 30'$. This choice guarantees that the total area of the projection equals the area of the sphere. A disadvantage, however, is that the parallels are considerably shortened at lower latitudes. This is visually expressed in a N–S stretching of the equatorial regions and can only be avoided by choosing a standard parallel at lower latitude.

The standard parallel can also be selected by quantitative analysis. Minimization of the overall scale distortion D_{ab} leads to a standard latitude $\phi_0 = 37.5°$ (Fig. 8.1). This map is shown in the introduction of the Directory (Fig. 4.1a). Minimization of the scale distortion over continental area D_{abs} yields a higher standard latitude ($\phi_0 = 43.0°$) as a result of the concentration of continental area in the middle latitudes. Since especially the elongation of the N–S distances in the lower latitudes is visually inconvenient it is recommended to opt for the first solution ($\phi_0 = 37.5°$). Although the total area is no longer correct the appearance of the continents is more pleasing than on Miller's equi-rectangular projection. Nevertheless the maximum distortion values are still too extreme to justify the use of this projection for world maps.

The cylindrical equidistant projection with two standard parallels is one of the oldest known projection systems. Ptolemaeus (second century AD) attributes it to Marinus of

Tyre (± 100 BC). Marinus' map of the ancient world is a cylindrical equidistant projection with standard latitude 36° (the parallel of Rhodes). Ptolemaeus considered this projection as unsatisfactory to map the *oikumene* (the inhabited world) because the meridians do not converge. He used the graticule only for the mapping of small areas with the middle parallel of the map represented at correct length. As such it appears in many editions of Ptolemaeus' *Geographia* which were printed in the fifteenth century as well as in the well-known Dutch atlases of the sixteenth century (Ortelius's *Theatrum orbis terrarum*, 1570; De Jode's *Speculum orbis terrarum*, 1578; Mercator's *Atlas*, 1595).

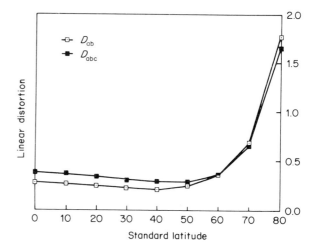

Fig. 8.1   Linear distortion as a function of standard latitude for the cylindrical equidistant projection.

Transformation formulas

$$x = R\lambda \cos \phi_0 \qquad\qquad y = R\phi \qquad\qquad \phi_0 = \text{standard latitude}$$

Distortion characteristics

Miller's equi-rectangular projection ($\phi_0 = 50° 30'$):

$D_{ar} = 0.50 \qquad D_{an} = 20.2 \qquad D_{ab} = 0.25$
$D_{arc} = 0.59 \qquad D_{anc} = 21.8 \qquad D_{abc} = 0.30$

Cylindrical equidistant projection with $\phi_0 = 37.5°$:

$D_{ar} = 0.43 \qquad D_{an} = 15.1 \qquad D_{ab} = 0.21$

Cylindrical equidistant projection with $\phi_0 = 43.0°$:

$D_{arc} = 0.57 \qquad D_{an} = 19.1 \qquad D_{ab} = 0.28$

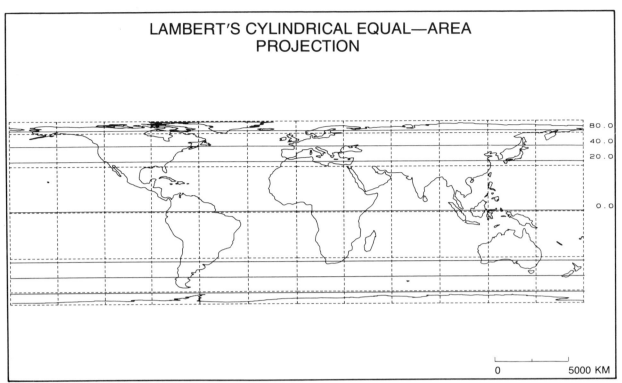

LAMBERT'S CYLINDRICAL EQUAL—AREA
PROJECTION

80.0
40.0
20.0

0.0

0   5000 KM

Lines of equal angular distortion

8.2   Cylindrical equal-area projection

The normal aspect cylindrical projections are characterized by a rectangular graticule where all parallels have the same length as the two correctly represented standard parallels. The general transformation formulas are given by

$$x = R\lambda \cos \phi_0$$

$$y = f(\phi)$$

where ϕ_0 = latitude of the standard parallels.

The properties of the projection are determined by the function $y = f(\phi)$. For the cylindrical equal-area projection this function is

$$y = R\frac{1}{\cos \phi_0} \sin \phi$$

Different cylindrical equal-area projections are obtained by proper choice of the standard parallel. The best known cylindrical equal-area projections are those of Lambert, Behrmann and the recently much-publicized Peters projection.

Transformation formulas

$$x = R\lambda \cos \phi_0$$

$$y = R\frac{1}{\cos \phi_0} \sin \phi$$

where ϕ_0 = latitude of the standard parallels.

8.2.1   *Cylindrical equal-area projection with one standard parallel*

Other name(s):   Lambert's cylindrical equal-area projection

Author(s):   Archimedes

Description

Lambert's projection can be considered as a limiting case of the cylindrical equal-area projection where the two standard parallels coincide with the equator. Although named after Lambert the projection is thought to have been developed by Archimedes. As a result of the equal-area property it is characterized by an enormous N–S compression at higher latitudes. This is reflected in a considerable mean angular distortion over the continental area. The projection is unsuited for world maps.

Distortion characteristics

$D_{ar} = 0.00$	$D_{an} = 30.9$	$D_{ab} = 0.55$
$D_{arc} = 0.00$	$D_{anc} = 37.8$	$D_{abc} = 0.77$

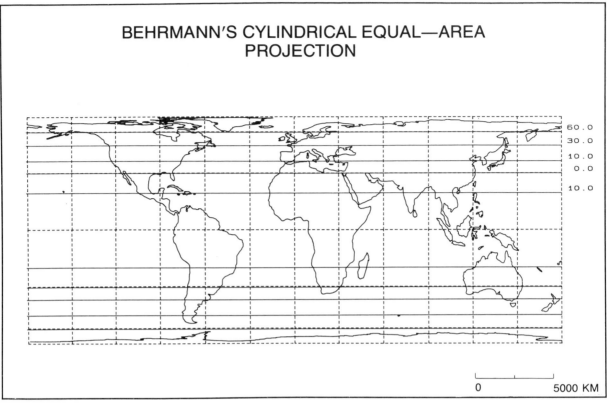

BEHRMANN'S CYLINDRICAL EQUAL—AREA PROJECTION

60.0
30.0
10.0
0.0
10.0

0   5000 KM

Lines of equal angular distortion

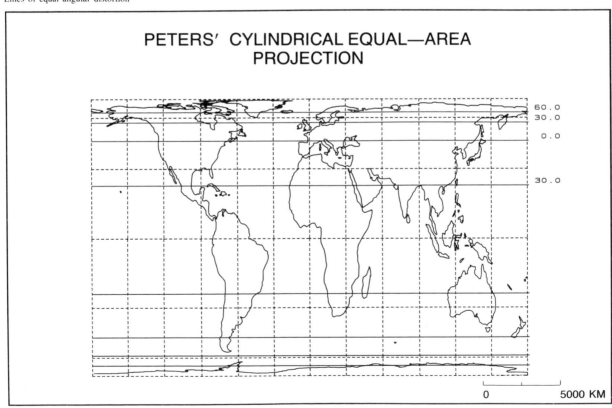

PETERS' CYLINDRICAL EQUAL—AREA PROJECTION

60.0
30.0
0.0

30.0

0   5000 KM

Lines of equal angular distortion

8.2 Cylindrical equal-area projection (CONTINUED)

8.2.2 Cylindrical equal-area projection with two standard parallels ($\phi_0 = \pm 30°$)

Other name(s):	Behrmann's projection
Author(s):	Walter Behrmann (1909)

Description

W. Behrmann (1909) made a comparative study of the distortion characteristics of some equal-area projections. He calculated a weighted mean of the angular distortion ('Durchschnittswinkelverzerrung') by drawing lines of equal distortion on the map and measuring the area between each two adjacent lines (Section 3.1.1). He calculated this parameter for a number of equal-area projections and obtained the lowest value for a cylindrical equal-area projection with standard latitude 30°. This is confirmed by the value of the distortion parameter D_{an}, which has the lowest value of all projections treated in this directory. Minimization of this parameter leads to a standard latitude of 27.5°, a value which is very close to Behrmann's 30°. Minimization of the mean angular distortion over continental area gives exactly 30° as standard latitude (Fig. 8.2).

In spite of this low mean angular distortion the projection is less suited for world maps than certain non-cylindrical equal-area projections. The distortion on a cylindrical projection increases with the distance from the standard parallel which leads to extreme scale distortions near the edges of the map.

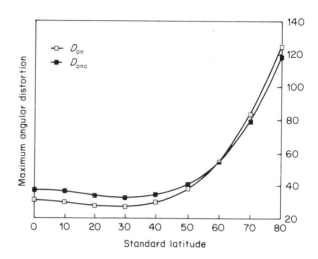

Fig. 8.2 Maximum angular distortion as a function of standard latitude for the cylindrical equal-area projection.

Distortion characteristics	$D_{ar} = 0.00$	$D_{an} = 26.8$	$D_{ab} = 0.44$
	$D_{arc} = 0.00$	$D_{anc} = 32.5$	$D_{abc} = 0.62$

8.2.3 Cylindrical equal-area projection with two standard parallels ($\phi_0 = \pm 45°$)

Other name(s):	Gall's orthographic projection/Peters's projection
Author(s):	James Gall (1885)

Description

The German historian A. Peters emphasized that on traditional projections (especially Mercator's projection) the areas in the middle and higher latitudes are generally represented at a larger scale than the equatorial areas. He subsequently developed an equal-area projection which, although designated as a revolutionary new projection, appeared to be a cylindrical equal-area projection with standard parallels at ±45°. Gall published such a cylindrical equal-area projection already in 1885. Many papers questioned the originality and the pretended qualities Peters attributed to his own projection (Kaiser, 1974; Maling, 1974; Loxton, 1985).

The selection of standard parallels at the relatively high latitude of 45° leads to an enormous N–S stretching of the equatorial areas. Surprisingly the projection is very popular with development agencies as a tool to illustrate the areal importance of the developing countries. The projection is completely unsuitable for general purpose world maps because of the extreme scale distortions.

Distortion characteristics	$D_{ar} = 0.00$	$D_{an} = 33.0$	$D_{ab} = 0.45$
	$D_{arc} = 0.00$	$D_{anc} = 36.9$	$D_{abc} = 0.57$

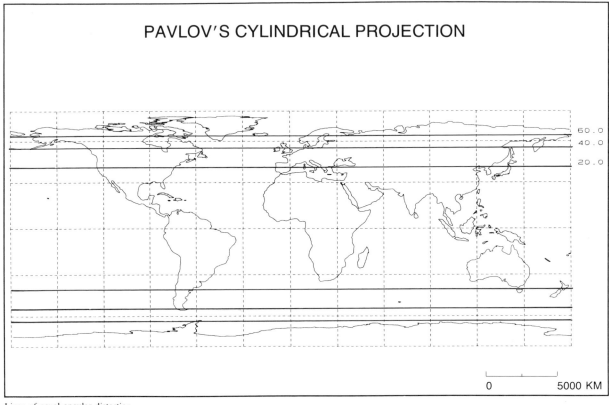

Lines of equal angular distortion

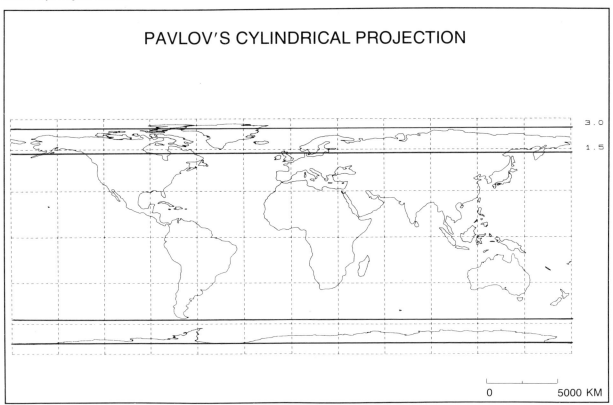

Lines of equal areal distortion

8.3 Cylindrical projection (Pavlov)

Author(s):   A. A. Pavlov

Description

Pavlov applied Urmaev's construction principle for cylindrical projections to develop a graticule that has totally different distortion characteristics than the projection proposed by Urmaev himself (Urmaev's cylindrical projection). This proves that Urmaev's theory allows for the development of a great variety of cylindrical projections depending on the imposed constraints. The transformation formulas of Pavlov's projection (Maling, 1960) differ from those of Urmaev by the values of the coefficients.

The distortion parameters are comparable with those obtained for projections of the pseudocylindrical class which give up the equal-area property to compensate the angular distortion. This class of projections is very popular in contemporary atlas cartography for general-purpose maps.

Pavlov's projection has the disadvantage of stretching the E–W distances in the higher latitudes. This is reflected in the relatively high values of the distortion parameters over continental areas. Therefore the projection is less suited for world maps than the above-mentioned pseudocylindrical projections which have more balanced distortion patterns.

Transformation formulas

$$x = R\lambda$$

$$y = R\left(a_0\phi + \frac{a_2}{3}\phi^3 + \frac{a_4}{5}\phi^5\right)$$

where $a_0 = 1.0000$, $a_2 = -0.1531$ and $a_4 = -0.0267$.

Distortion characteristics

$D_{ar} = 0.31$	$D_{an} = 21.5$	$D_{ab} = 0.33$
$D_{arc} = 0.43$	$D_{anc} = 26.8$	$D_{abc} = 0.45$

MERCATOR'S PROJECTION

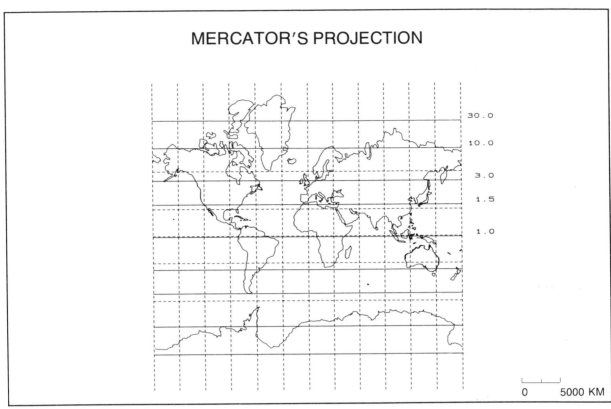

30.0

10.0

3.0

1.5

1.0

0   5000 KM

Lines of equal areal distortion

8.4 Cylindrical conformal projection (Mercator)

Other name(s): Mercator projection

Author(s): Gerhard Mercator (1569)

Description

Mercator's projection is a conformal projection. This implies that the particular scales in any point on the map are equal in all directions. Since Mercator's projection represents the equator in correct length the parallels are considerably stretched in the higher latitudes. Therefore to ensure conformality the spacing of the parallels increases rapidly. The pole itself cannot be represented. The enormous areal exaggeration makes the projection unsuited for general purpose world maps. This has not always been realized. Yet Mercator developed his projection for navigational purposes and therefore called it 'ad usum navigantium' (1569). The Mercator chart has the important property of representing the loxodrome as a straight line. The map was not included in the 1595 first edition of Mercator's *Atlas*, which was published after his death by his son Rumold. It is only in the later editions, from 1637, that the world map on a double stereographic projection was replaced by a map on Mercator's conformal projection. During the second half of the seventeenth and the largest part of the eighteenth century the projection was missing in the well-known atlases. In the first half of the nineteenth century it reappeared in the famous German 'Stieler'-atlas (first edition 1817–23). The misuse of Mercator's projection in atlas cartography and especially in school atlases still leads to a lot of geographical misconceptions.

Transformation formulas

$$x = R\lambda$$

$$y = R \ln \tan \left(\frac{\pi}{4} + \frac{\phi}{2} \right)$$

Distortion characteristics

$D_{ar} = 3.40$	$D_{an} = 0.0$	$D_{ab} = 0.55$
$D_{arc} = 5.45$	$D_{anc} = 0.0$	$D_{abc} = 0.77$

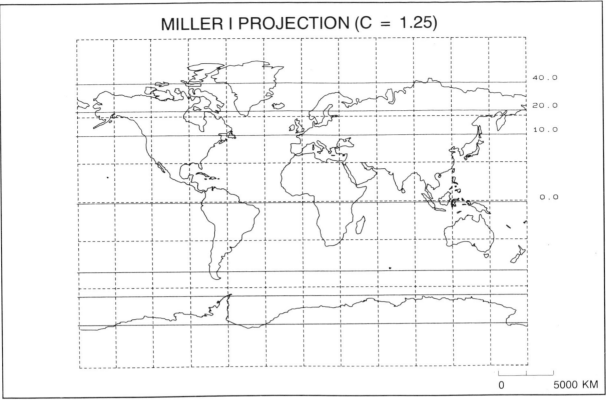

Lines of equal angular distortion

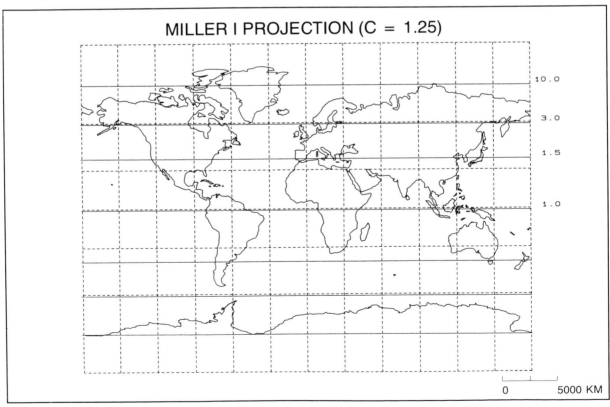

Lines of equal areal distortion

8.5 Cylindrical projection (Miller)

Other name(s): Miller I/Miller II projection

Author(s): O. M. Miller (1942)

Description:

The conformal property of Mercator's projection (cylindrical conformal projection) leads to an extreme areal exaggeration at higher latitudes. This makes the projection totally unsuited for general-purpose world maps. The other extreme, the cylindrical equal-area projection, which combines an elongation of the N–S distances in the lower latitudes with a N–S compression of the polar areas, is also not appropriate for general use in atlas cartography. The cylindrical equidistant projection occupies an intermediate position between these two extremes although it is still characterized by a considerable shearing in the higher latitudes. Therefore, other cylindrical projections were developed with intermediate properties. These projections reduce in the first place the variability of scale that might lead to excessive shearing in a certain direction. They also give up absolute conformality to avoid extreme areal exaggeration. Gall's cylindrical stereographic projection is one of the oldest and best known projections to accomplish this. It is often used in British atlases.

In 1942 O. M. Miller published two projections which may be considered as alternatives for Gall's projection. He introduced a constant C in the transformation formulas of Mercator's projection. For $C = 1$ the Mercator projection itself is obtained. Increasing the value of C allows for a reduction of the areal distortion at the cost of a growing angular distortion. Miller's cylindrical projections are characterized by $C = 1.25$ (Miller I projection) and $C = 1.5$ (Miller II projection).

Miller's first projection is frequently used in American atlases. Compared with Gall's stereographic projection we notice that the latter has a considerably lower mean areal distortion for a slightly higher mean angular distortion. This is mainly the result of Gall's choice of a standard parallel at $45°$. Miller's projections represent the equator at correct length.

Miller's second projection has a higher value for the constant C. Therefore the areal distortion is lower than for Miller's first projection at the expense of a higher angular distortion. The graticule of this projection is almost identical to Braun's cylindrical stereographic projection and has therefore similar distortion characteristics. Miller's second projection never reached the popularity of his first.

Together with the cylindrical stereographic projection (Gall, BSAM, Braun) Miller's projections belong to the group of projections which approximate conformality rather than equivalence. Therefore they have the same application area. They are especially suited for vector representations (ocean currents, wind gradients, etc.).

Transformation formulas:

$$x = R\lambda \qquad y = CR\ln\,\tan\left(\frac{\pi}{4} + \frac{\phi}{2C}\right)$$

Miller I: $C = 1.25$
Miller II: $C = 1.50$

Distortion characteristics

| *Miller I projection:* | $D_{ar} = 1.23$ | $D_{an} = 7.6$ | $D_{ab} = 0.38$ |
| | $D_{arc} = 1.79$ | $D_{anc} = 9.9$ | $D_{abc} = 0.52$ |

| *Miller II projection* | $D_{ar} = 0.93$ | $D_{an} = 10.8$ | $D_{ab} = 0.34$ |
| | $D_{arc} = 1.33$ | $D_{anc} = 13.8$ | $D_{abc} = 0.47$ |

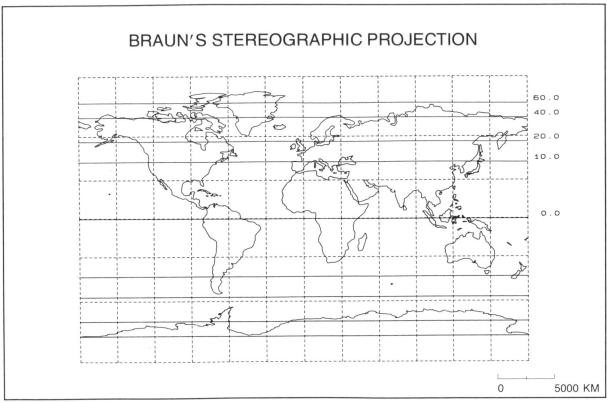

Lines of equal angular distortion

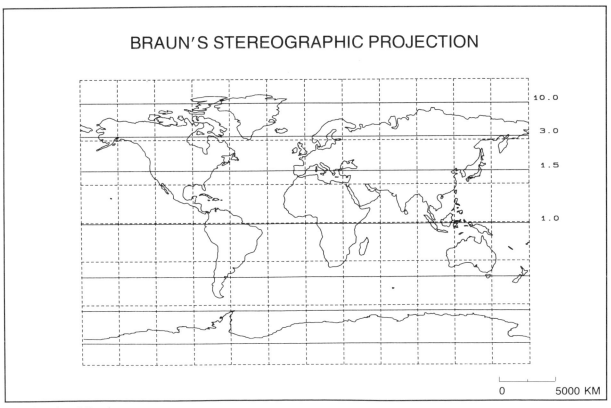

Lines of equal areal distortion

8.6   Cylindrical stereographic projection

This projection system results from the stereographic projection of the globe on a cylinder that intersects the globe along two standard parallels which are symmetric about the equator.

Projections of this type differ from one another in the choice of the standard parallels. They are neither conformal nor equal-area. Some of them offer a good compromise between areal and angular distortion. Three projections are well known in cartographic literature and practice: Braun's perspective projection, the Bolshoi Sovietsky Atlas Mira (BSAM) projection and especially Gall's stereographic projection.

Transformation formulas

$$x = R\lambda \cos \phi_0$$

$$y = R(1 + \cos \phi_0) \tan \frac{\phi}{2}$$

$$\phi_0 = \text{latitude of standard parallels}$$

8.6.1   *Cylindrical stereographic projection with one standard parallel*

Other name(s):   Braun's perspective cylindrical projection

Author(s):   P. Braun

Description

The projection is a special case of the cylindrical stereographic projection with the cylinder touching the sphere along the equator instead of intersecting it (Steers, 1970). This implies that the two standard parallels coincide with the equator ($\phi_0 = 0$).

The projection shows a considerable E–W stretching in the higher latitudes. The mean areal distortion over the continental area is also relatively high. The graticule is almost identical to Miller's second projection (Miller II). Both projections are less suitable for world maps. The cylindrical stereographic projections with two standard parallels (Gall, BSAM) are better alternatives among the group of cylindrical projections which give up the absolute conformality to avoid extreme scale and areal distortion.

Distortion characteristics

$D_{ar} = 0.96$	$D_{an} = 10.3$	$D_{ab} = 0.35$
$D_{arc} = 1.36$	$D_{anc} = 13.3$	$D_{abc} = 0.47$

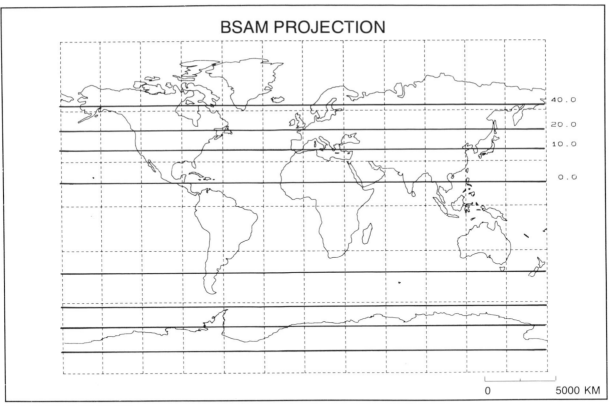

Lines of equal angular distortion

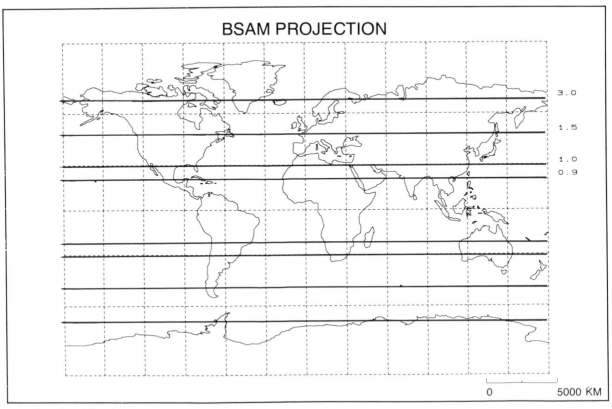

Lines of equal areal distortion

8.6 Cylindrical stereographic projection (CONTINUED)

8.6.2 *Cylindrical stereographic projection with two standard parallels* ($\phi_0 = \pm 30°$)

Other name(s): Bolshoi Sovietsky Atlas Mira projection/
BSAM projection

Description

This projection was used in the first volume of the Bolshoi Sovietsky Atlas Mira for world distribution maps and is therefore often called the BSAM projection (Maling, 1960). The choice of the standard parallels at 30° leads to more favorable distortion patterns than for Braun's projection. This is reflected in a lower value for all calculated distortion parameters.

Distortion characteristics

$$D_{ar} = 0.73 \qquad D_{an} = 9.0 \qquad D_{ab} = 0.28$$
$$D_{arc} = 1.03 \qquad D_{anc} = 11.6 \qquad D_{abc} = 0.38$$

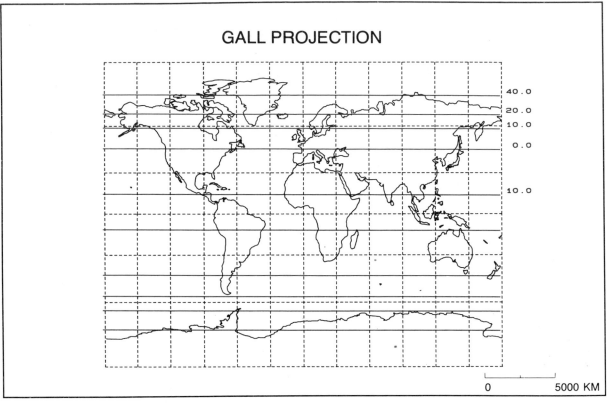

GALL PROJECTION

40.0
20.0
10.0
0.0

10.0

0   5000 KM

Lines of equal angular distortion

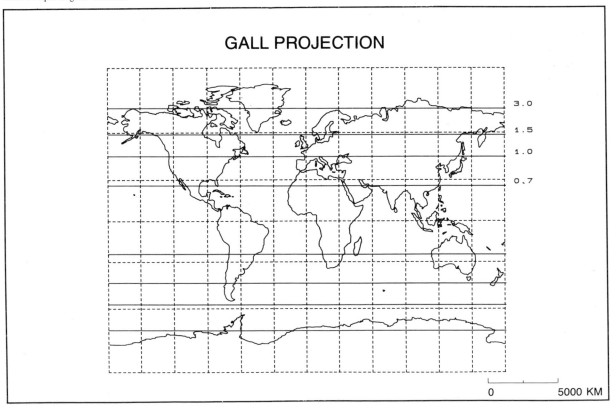

GALL PROJECTION

3.0
1.5
1.0

0.7

0   5000 KM

Lines of equal areal distortion

8.6 Cylindrical stereographic projection (CONTINUED)

8.6.3 *Cylindrical stereographic projection with two standard parallels* ($\phi_0 = \pm 45°$)

Other name(s):	Gall's (stereographic) projection
Author(s):	James Gall (1885)

Description

Among the three cylindrical stereographic projections described Gall's is definitely the best known. It was developed by the end of the nineteenth century by Rev. James Gall and is characterized by two standard parallels at 45°. It is often used in British atlases. Compared with the BSAM projection the E–W stretching, as well as the area exaggeration in the higher latitudes, is less pronounced as the result of the selection of the standard parallel at a higher latitude. The global scale distortion is less for the BSAM projection. When, on the other hand, the scale distortion is integrated over continental areas Gall's projection has a smaller mean scale distortion because its standard parallel is better centered with respect to the large continental masses of the Northern Hemisphere.

In both the *Times Atlas* and the *Oxford Atlas* a modification of Gall's projection is used for a number of world maps. The modification for the *Times Atlas*, which is called the 'The Times' projection, was developed by Moir, the one for the *Oxford Atlas*, simply called a 'Modified Gall' projection, by Guy Bomford. On both projections the spacing of the parallels remains unaltered, but the meridians are slightly curved to decrease the scale exaggeration near the poles. Therefore these projections form part of the pseudocylindrical class. The difference between the two modifications is the way in which the meridians are curved. Snyder (1977) gives approximating formulas to the x-coordinates of both projections.

Distortion characteristics

$D_{ar} = 0.70$	$D_{an} = 10.6$	$D_{ab} = 0.29$
$D_{arc} = 0.89$	$D_{anc} = 12.5$	$D_{abc} = 0.36$

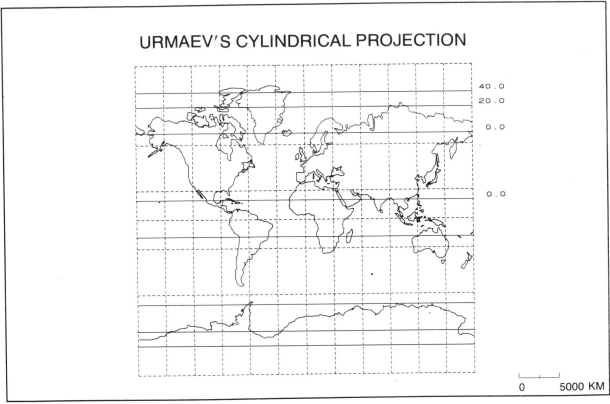

URMAEV'S CYLINDRICAL PROJECTION

40.0
20.0
0.0

0.0

0   5000 KM

Lines of equal angular distortion

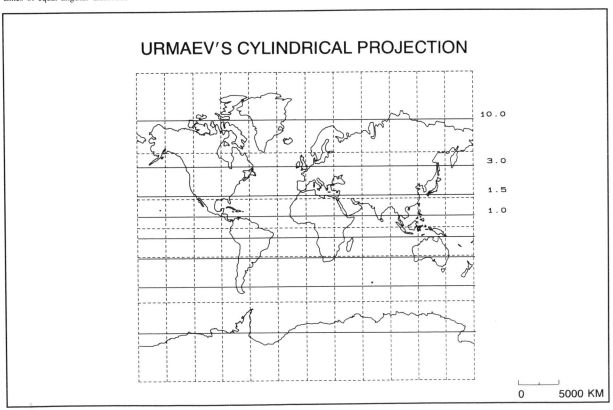

URMAEV'S CYLINDRICAL PROJECTION

10.0

3.0

1.5

1.0

0   5000 KM

Lines of equal areal distortion

8.7 Cylindrical projection (Urmaev)

Other name(s): Urmaev III projection

Author(s): N. A. Urmaev (1947)

Description

N. A. Urmaev belongs to the Russian school of mathematical cartography (Maling, 1960). He developed a general cylindrical projection system which allows us to construct an infinite number of cylindrical projections with different distortion characteristics.

For cylindrical projections the linear scale along the meridians h is a function of the latitude and may be expressed by the polynomial

$$h = R(a_0 + a_2\phi^2 + a_4\phi^4 + \ldots)$$

On the other hand, h can also be expressed as a function of the areal distortion or the maximum angular distortion. This implies that when certain distortion constraints are imposed upon different parallels the corresponding h can be calculated and the coefficients a_0, a_2, a_4, ... obtained by solving the following system of equations

$$h_1 = R(a_0 + a_2\phi_1^2 + a_4\phi_1^4 + \ldots a_{2(n-1)}\phi_1^{2(n-1)})$$

$$h_2 = R(a_0 + a_2\phi_2^2 + a_4\phi_2^4 + \ldots a_{2(n-1)}\phi_2^{2(n-1)})$$

$$\vdots$$

$$h_n = R(a_0 + a_2\phi_n^2 + a_4\phi_n^4 + \ldots a_{2(n-1)}\phi_n^{2(n-1)})$$

The ordinate y is given by

$$y = \int_0^\phi h\, \mathrm{d}\phi$$

Integration after substitution of the polynomial expression for h gives

$$y = R\left(a_0\phi + \frac{a_2}{3}\phi^3 + \frac{a_4}{5}\phi^5 + \ldots\right)$$

In practice the number of coefficients is generally restricted to three (a_0, a_2, a_4).

Urmaev developed in this way a projection which has been used for the political map of the world in the *Atlas Ofitsera* (1947). It is known as the Urmaev III projection (Maling, 1960). The distortion pattern of the projection is characterized by four parallels along which no angular distortion occurs ($\phi = \pm 20°$, $\phi = \pm 65°$). This illustrates how Urmaev widened the concept of the cylindrical projection. In contrast with the familiar cylindrical projections, angular distortion now takes place along the parallel which is represented in correct length (in this case the equator) while, on the other hand, the distortion becomes zero on four other parallels. Practically the whole continental area is situated within the isoline of 20°. The angular distortion parameters have a very low value. On the other hand the projection shows a considerable areal exaggeration in the middle and higher latitudes (compare the area of Scandinavia with the area of the rest of Europe).

Other Russian cartographers constructed similar cylindrical projections starting from Urmaev's principle. Maling (1960) for instance, in his article on Russian map projections, describes the projections of Pavlov (also in this directory) and Shabanova.

Transformation formulas

$$x = R\lambda \qquad y = R\left(a_0\phi + \frac{a_2}{3}\phi^3\right)$$

where $a_0 = 0.9281$ and $a_2 = 1.1143$.

Distortion characteristics

$D_{ar} = 1.84$	$D_{an} = 5.4$	$D_{ab} = 0.51$
$D_{arc} = 2.61$	$D_{anc} = 6.3$	$D_{abc} = 0.67$

REFERENCES

Arden-Close, Sir Charles (1952), A forgotten pseudo-zenithal projection, *Geographical Journal*, **118**, 237.

Baar, E. J. (1947), The manipulation of projections for world maps, *Geographical Review*, **37**, 112–120.

Behrmann, W. (1909), Zur Kritik der flächentreuen Projektionen der ganzen Erde und einer Halbkugel, *Sitzungsberichte der Königlich Bayerischen Akademie der Wissenschaften, Mathematisch-physikalische Klasse*, Munich (13), 19–48.

Behrmann, W. (1910), Die beste bekannte flächentreue Projektion der ganzen Erde, *Petermanns Geographische Mitteilungen*, **56**, II (3), 141–4.

Boggs, S. W. (1929), A new equal-area projection for world maps, *Geographical Journal*, **73**, 241–5.

Briesemeister, W. (1953), A new oblique equal-area projection, *Geographical Review*, **43**, 260–1.

Craster, J. E. E. (1929), Some equal-area projections of the sphere, *Geographical Journal*, **74**, 471–4.

Driencourt, L., and Laborde, J. (1932), *Traité des projections des cartes géographiques à l'usage des cartographes et des géodésiens*, 4 fascicules, Hermann et Cie, Paris.

Eckert, M. (1906), Neue Entwürfe für Erdkarten, *Petermanns Geographische Mitteilungen*, **52** (V), 97–109.

Eckert-Greifendorff, M. (1935), Eine neue flächentreue (azimutaloide) Erdkarte, *Petermanns Geographische Mitteilungen*, **81**, 190–2.

Gall, Rev. James (1885), Use of cylindrical projections for geographical, astronomical, and scientific purposes, *Scottish Geographical Magazine*, **1**, 119–23.

Gambier, G. (1975), *Notions sur les représentations planes de la terre*, Institut Géographique National, Paris, p. 102.

Grafarend, E., and Niermann, A. (1984), Beste echte Zylinderabbildungen, *Kartographische Nachrichten*, **34** (3), 103–7.

Hammer, E. (1892), Über die Planisphäre von Aitow und verwandte Entwürfe, insbesondere neue flächentreue ähnlicher Art, *Petermanns Geographische Mitteilungen*, **38**, 85–7.

Hammer, E. (1900), Unechtzylindrische und unechtkonische flächentreue Abbildungen, *Petermanns Geographische Mitteilungen*, **46**, 42–6.

Horn, W. (1961), Zur Geschichte der Atlanten, *Kartographische Nachrichten*, **11** (1), 1–8.

Hoschek, J. (1969), *Mathematische Grundlagen der Kartographie*, Bibliographisches Institut, Mannheim–Zürich, p. 167.

Hsu, M.-L. (1981), The role of projections in modern map design, in L. Guelke (ed.), *Maps in Modern Geography, Geographical Perspectives on the New Cartography*, Cartographica, Monograph 27, **18** (2), 151–86.

Kaiser, A. (1974), Die Peters-Projektion, *Kartographische Nachrichten*, **24** (1), 20–5.

Keuning, J. (1955), The history of geographical map projections until 1600, *Imago Mundi*, **12**, 1–24.

Lopez, B. (1987), *Arctic Dreams*, Bantam Books, Toronto–New York, p. 417.

Loxton, J. (1985), The Peters phenomenon, *The Cartographic Journal*, **22** (2), 106–8.

Maling, D. H. (1960), A review of some Russian map projections, *Empire Survey Review*, **15** (115,116,117), 203–15, 255–66, 294–303.

Maling, D. H. (1968), The terminology of map projections, *International Yearbook of Cartography*, **8**, 11–65.

Maling, D. H. (1973), *Coordinate Systems and Map Projections*, George Philip & Son, London, p. 255.

Maling, D. H. (1974), Peters' Wunderwerk, *Kartographische Nachrichten*, **24** (4), 153–6.

McBryde, F. W. and Thomas, P. D. (1949), *Equal-Area Projections for World Statistical Maps*, U.S. Coast and Geodetic Survey Spec. Pub. No. 245, Washington, p. 44.

Mercator, G. (1963), *Atlas sive cosmographicae meditationes de fabrica mundi et fabrica figura*, (1595), Culture et Civilisation, Brussels, p. 32.

Miller, O. M. (1942), Notes on cylindrical world map projections, *Geographical Review*, **32**, 424–30.

Miller, R. (1949), An equi-rectangular map projection, *Geography*, **34**, 196–201.

Nell, A. M. (1890), Äquivalente Kartenprojektionen, *Petermanns Geographische Mitteilungen*, **36**, 93–8.

O'Keefe, J. A. and Greenberg, A. (1977), A note on the van der Grinten projection of the whole earth onto a circular disk, *The American Cartographer*, **4** (2), 127–32.

Ortelius, A. (1964), *Theatrum orbis terrarum*, (1570), facsimile edition, first series, vol. III, Meridian, Amsterdam.

Peters, A. (1975), Wie man unsere Weltkarten der Erde ähnlicher machen kann, *Kartographische Nachrichten*, **25** (5), 173–83.

Peters, A. (1978), Über Weltkartenverzerrungen und Weltkartenmittelpunkte, *Kartographische Nachrichten*, **28** (3), 106–13.

Peters, A. (1982), Zur Theorie der Entfernungsverzerrung, *Kartographische Nachrichten*, **32** (4), 132–4.

Putnins, R. V. (1934), Jaunas projekcijas pasaules kartem, *Geografiski raksti*, Folia Geographica III, IV, Riga, 180–209.

Reignier, F. (1957), *Les systèmes de projection et leurs applications à la géographie, à la cartographie, à la navigation, à la topométrie, etc …*, Publications techniques de l'Institut Géographique National, Paris.

Richardus, P. and Adler, R. K. (1972), *Map Projections*, North-Holland Publishing Company, Amsterdam, p. 174.

Robinson, A. H. (1951), The use of deformational data in evaluating world map projections, *Annals of the Association of American Geographers*, **41** (1), 58–74.

Robinson, A. H. (1974), A new map projection: its development and characteristics, *International Yearbook of Cartography*, **14**, 145–55.

Shirley, R. W. (1984), *The Mapping of the World*, Holland Press Cartographica Series, 9, London, p. 669.

Snyder, J. P. (1977), A comparison of pseudocylindrical map projections, *The American Cartographer*, **4** (1), 59–81.

Snyder, J. P. (1979), Projection Notes, *The American Cartographer*, **6** (1), 81.

Snyder, J. P. (1982), *Map Projections Used by the U.S. Geological Survey*, Geological Survey Bulletin 1532, United States Government Printing Office, Washington, p. 313.

Snyder, J. P. (1985), *Computer-Assisted Map Projection Research*, US Geological Survey Bulletin 1629, United States Government Printing Office, Washington, p. 157.

Steers, J. A. (1970), *An introduction to the Study of Map Projections*, University of London Press, p. 294.

Tissot, A. (1881), *Mémoire sur la représentation des surfaces et les projections des cartes géographiques*, Gauthier Villars, Paris, p. 171.

Tobler, W. R. (1962), A classification of map projections, *Annals of the Association of American Geographers*, **52**, 167–75.

Tobler, W. R. (1968), Geographic area and map projections, in B. J. L. Berry and D. F. Marble (eds.), *Spatial Analysis*, Prentice Hall, 78–90.

Van der Grinten, A. J. (1904), Darstellung der ganzen Erdoberfläche auf einer kreisförmigen Projektionsebene, *Petermanns Geographische Mitteilungen*, **50**, 155–9.

Wagner, K. (1962), Kartografische Netzentwürfe, 2nd edn., *Bibliographisches Institut*, Mannheim, 303.

Werenskiold, W. (1944), A class of equal-area projections, *Det Norske Videnskaps-Akademi Oslo, Matematisk-naturvidenskapelig Klasse* (11).

Winkel, O. (1928), Übersicht der Gradnetzkombinationen, *Petermanns Geographische Mitteilungen*, **74**, 201–4.

Winkel, O. (1939), 25 Jahre Neue Netzkombinationen, *Petermanns Geographische Mitteilungen*, **85**, 278–80.

INDEX